Marine Fisheries Ecosystem

Its quantitative evaluation and management

Taivo Laevastu and Herbert A Larkins

Northwest and Alaska Fisheries Center
and
Northwest Regional Office
National Marine Fisheries Service, NOAA
Seattle, Washington, USA

Fishing News Books Ltd
Farnham · Surrey · England

British Library CIP Data
Laevastu, Taivo
Marine fisheries ecosystem
1. Fishery management – Mathematical models
I. Title   II. Larkins, Herbert A
333.95′6′17   SH334

ISBN 0 85238 116 6

Published by Fishing News Books Ltd
1 Long Garden Walk, Farnham, Surrey

Printed in Great Britain by
Page Bros (Norwich) Ltd, Norwich

Typesetting by Traditional Typesetters Ltd, Chesham

Contents

Figures

Tables

Preface

There is a natural evolution in the development of science. Creativity, inhibited by the absence of appropriate instrumentation, may be unleashed with explosive force when new tools become available. In many, if not most instances, the tool makers themselves are far removed from the impacts of their innovation. Such a case is clearly reflected in the advances in many diverse fields that have been made possible through the technological breakthroughs that have permitted the creation of high powered high capacity computers.

Scientists finally have the tools to allow them the opportunity to generate and test theories regarding the complex interrelationship between the physical, chemical and biological components of an ecosystem. During the past decade ecosystem modeling has emerged as a major new direction of study. Models have been generated to describe the tundra, grasslands, and estuaries, *etc*. We have also begun, after years of frustrating anticipation, to model the ecosystems involving exploited species.

It must be clearly understood that the realization of the potential for ecosystems models to provide a basis for the rational management of our living marine resources remains in the future. This book, which summarizes the research that has been accomplished at the Northwest and Alaska Fisheries Center, represents a building block which we hope will be useful in the construction of an edifice that will permit us to predict the consequences of fishing activities and other environmental perturbations, both natural and man-made.

Our realization of the complexity of the ecosystems is not new and the use of 'single species' models by many fisheries scientists resulted not from naiveté on their part, but simply from the absence of tools to permit the use of more complex models. Those of us who labored in the age before chip technology well remember the hours spent pounding mechanical calculators to accomplish an analysis of variance, co-variance, or plot a von Bertalanffy growth curve. We 'knew', and continue to 'know', that the exploitation of one species has an impact on other elements in the system. The intuitive belief, either unsupported or poorly supported by data, that minke whale populations in the Southern Ocean have increased because of the removal of other baleen whales, that the carrying capacity for the Northern fur seals in the Bering Sea is less today than it was before the advent of major bottom fishing operations, that the large take of Pacific pollock is influencing the recruitment and growth rates of other fishes, is a caveat understood and acted upon in the management of these stocks. The absence of an ability to quantify these impacts, however, forces the resource manager to implement conservative approaches to management, which in most cases means that lower quotas than might be allowed with full knowledge, are established.

The reader of this volume must avoid the temptation of accepting any model as the real world or the results of the model as meaning more than that which results from both the data and the assumptions used in the model. The caution of Sir Napier Shaw, 'Every theory of the course of events in nature is necessarily based on some process of simplification and is to some extent therefore, a fairy tale.' should not be forgotten. The work that follows is neither a fairy tale nor a completed task. It is a summary presented with the hope that it will inspire others to diligently pursue the tasks of closing the large uncertainties and gaps in our understanding of how ecosystems work.

We must in the future have both better data, particularly in terms of the relationship between environmental factors and the survival of early life history stages of marine species, and better models that permit the examination of feed-back mechanisms. For example, while marine mammals eat fish, and serve as vectors for a number of parasites that may adversely affect fish, they also eat piscivorous species. The consequences of maintaining large marine mammal populations to exploitable populations of finfish remain undefined. Solutions to this and other critical problems involving the conservation and management of our marine resources including the impact of oil developments, incidental takes, marine pollution, *etc*, may very well be realized through the evolution and application of ecosystem models.

William Aron,
Center Director,
NWAFC,
Seattle, Washington

Foreword

Profound changes have occurred in marine fisheries in the last decade all over the world – many pelagic stocks have declined rather abruptly (*eg* herring in the North Atlantic, anchovy off Peru), the increase of catches of most species in high latitudes in the Northern Hemisphere has stopped, and major fisheries have shifted from one species to another. Many of these changes indicate that most major fishery resources in the world seem to be under intensive exploitation and some stocks have been overexploited. One of the most profound changes in the fisheries in 1970s was caused, however, by the extension of jurisdiction over fishery resources and their management by coastal states to 200 nautical miles from the coast. This change has shifted the management of major fishing grounds to individual coastal states, in contrast to earlier practices of major fishery resources management by intergovernmental (and international) councils. These international councils are still important in coordinating fisheries research and management.

The new management approach by coastal states requires that the fishery resources under their jurisdiction be managed on the basis of best available knowledge, *ie* knowledge of the magnitudes of the resources, how they fluctuate in space and time, how marine mammals affect them, and how they respond to changing fisheries. These requirements have also changed many aspects of applied fisheries research. It has been recognized that data collections from commercial vessels in fully-managed fisheries are often directed by management measures and do not necessarily fulfill some earlier resource survey purposes. Thus, there is a multiple need: to improve total resource surveys, extend these surveys to prefishery juveniles, and compliment these surveys with other indirect means, such as with quantitative ecosystem simulations.

It has been realized that single-species population dynamics models are no longer sufficient for many purposes in fisheries management for which they were used in the 1950s and 1960s. Furthermore, it has been realized that a fishery affects not only the target species, but also other nontarget species in the ecosystem via interspecies interactions (mainly predation). Thus, we need to take a synthetic look at the total marine ecosystem, utilizing all accumulated knowledge about it, and start to manage the marine fisheries on an ecosystem basis.

This book is a summary of experiences and results of ecosystem simulation in the Northwest and Alaska Fisheries Center (NWAFC) in Seattle, gained during the last four years. The approach taken in NWAFC is somewhat different from that of Andersen and Ursin (1977) – the only other known fisheries ecosystem (multispecies) model at present – partly because both approaches were developed separately, the main problems and conditions are different in the NE Pacific than in the North Sea, and the data availability is different in both areas.

This book is not intended 'to set concepts into concrete' as warned by Radovich (1976). The application of the ecosystem approach in fisheries is in its infancy: the future will surely bring changes and improvements to this approach. Therefore, the main purpose of this book is to share experiences and to further scientific-technological knowledge in some areas of the rapidly developing and changing fisheries science and its application.

1

Purpose and scope

The extension of fisheries jurisdiction by coastal states out to 200 miles from the coast brings most of the world fishery resources under the management of the coastal states. To manage these resources on sound national bases, one has to know the magnitudes of these resources, their distribution and behaviour in space and time, and their response to fishery and to proposed management options.

A need exists nearly everywhere for better information on the fishery resources and for a greater efficiency of data collection. Furthermore, a need for new analytical and prognostic techniques also exists, which would allow the evaluation of the management options before they are executed.

Total ecosystem simulation techniques promise a useful, new additional tool for the abovementioned tasks. Ecosystem simulation models can also be used for the determination of the magnitudes and periods of long-term fluctuations of the stocks caused by factors other than fishery (*eg* by environmental anomalies). We describe in this book our marine fisheries ecosystem simulation and results obtained from its application.

In the past a large part of fisheries science and its application has been based on the so-called single-species approach, *ie* considering the life history and biological processes of one species at a time and in a manner as though other species did not exist in the same area and interact in various ways with each other. Although attempts to consider the whole marine ecosystem as an integral unit have not been totally absent, progress in its quantitative modeling has been slow. However, we have learned about the advantages and shortcomings of single-species approaches and have begun to consider entire marine ecosystems in specific regions even though we are still interested primarily in knowing what is happening to the population of a single species.

This book is an attempt to summarize in quantitative terms the major processes of a marine ecosystem although emphasis has been placed on the fish biota and its relation to the environment. The main purpose is, however, to describe an approach to quantitative, numerical ecosystem simulation applicable to fisheries research and management purposes. The biological and ecological aspects of various processes are described only to the extent necessary to explain their application in simulation. The principles of large-scale simulation (modeling) presented here are based on experiences derived from not

1

only ecosystem simulation studies but also from a variety of large-scale meteorology and oceanography model studies. Large-scale simulation models use a multitude of modeling (simulation) techniques and are multipurpose, whereas models designed for special purposes tend to be simplifications of processes and conditions and often present only a caricature of one selected aspect of nature.

The applications of ecosystem approaches to fisheries problems are in their infancy and, therefore, many background and input data need to be developed and investigated. A selection has been made from the quantitative approaches presently available in the literature and suitable for use in ecosystem simulation, based on the availability of data and possibility of verification of results. There are few areas in science where there is a unanimous consensus in a given approach to, or a single solution to, a given problem. Numerous approaches are possible in the different subject areas described in this book, and different individuals (simulators) may prefer different techniques. An excellent, extensive fisheries multispecies model has been published and is in use in Denmark by Andersen and Ursin (1977). Other, slightly different models will surely follow (*eg* Balchen, 1980). A similar model development and succession has occurred in meteorology where the first numerical models were either barotropic and/or baroclinic, followed by primitive equation models which are lately competing with vectorial spherical harmonic models. The use and test of different approaches, including their comparison, is desirable for the advancement of the science and technology.

Traditionally, fisheries population dynamics has been number-based, *ie* based on considerations of the dynamics of the numbers of fish in a population. The technique presented here is a biomass-based approach because experience indicates that in data-poor areas biomass-based models are in many ways preferable to number-based approaches. Many approaches, however, apply with minor changes and/or conversions to either number- or biomass-based simulations.

Traditionally, most of the fish population dynamics models have been built on assumptions that a given phenomenon or an important relationship can be described by a mathematical equation. As pointed out by Radovich (1976), there is a tendency to forget that mathematical models of fish populations have not always been precise, detailed descriptions of real and complex biological phenomena – phenomena that are indeed difficult to model as many variables affect them. In this book some of these difficulties have been avoided by representing the essential processes by many relatively simple, and empirically tested formulas, which we attempt to explain in detail.

This book is not meant for the indoctrination of anyone in the use of the described models. It is merely a discussion and documentation of methods used in the formulation and evaluation of a holistic, fisheries-orientated ecosystem simulation and its evaluation. The references in this book are kept to a minimum and given usually in cases where detailed descriptions are necessary and/or available or where a controversial subject needs validation.

It is possible to organize the materials presented in this book in a number of different ways. Thus, a few sentences of explanation of the organization of the contents of this book are called for.

Chapter 2 presents a brief description of the marine ecosystem and some of its properties. A brief history of the past developments of quantitative evaluation of this system is given, whereby the general properties of single-species population dynamics approaches and their shortcomings are briefly described. The objectives of the holistic marine ecosystem simulation on the main principles followed in this task are also explained in this chapter.

Chapter 3 reviews difficulties encountered in quantitative evaluation of marine ecosystem which start with primary production and follows the fate of organic matter through the food chains. This chapter explains the reasons why we start the ecosystem simulation from the upper end of the food pyramid – *ie* we use a 'downside up' ecosystem. Furthermore, this chapter describes briefly the difficulties encountered in customary number-based fisheries dynamics models, and presents the principles of biomass-based model, used in this work.

Chapter 4 presents methods and formulas for computation of essential biomass parameters, their distribution with age in fish population, and changes in these computed parameters as affected by processes within fish populations, such as changes in recruitment and fishing.

Some users of this book might wish to turn after Chapter 4 to Chapter 6 where the formulas used in our simulation are assembled in a computationally logical sequence. Other users, especially those requiring some background in quantitative presentation of various processes in the fish-oriented ecosystem, might wish to consult Chapter 5 first. Chapter 5 is limited to the evaluation of quantitative simulation of those processes in fish populations which are essential in biomass-based ecosystem simulation.

Chapter 7 describes the role of plankton and benthos in the fish ecosystem.

Chapter 8 lists the essential inputs for fish ecosystem simulation and describes the method used for quantitative computation of biomasses in a given ecosystem – *ie* evaluation of the fish resources. The following Chapter 9 gives results of equilibrium biomass computations in the eastern Bering Sea.

Chapter 10 describes the environment-biota interaction simulation with examples, and Chapter 11 gives examples of magnitudes and periods of fluctuations in marine fish ecosystem, especially rate of change of individual species biomasses with time. The consumption of biota by marine mammals is summarized in Chapter 12.

Verification and validation of ecosystem simulations are described in Chapter 13. Chapters 14 to 16 describe the use of the ecosystem simulation in modern fisheries management tasks.

The symbols used in the formulas are listed in Chapter 17. As the numerical simulation concerns scientists from different disciplines, it was found necessary to add explanation of terms as Chapter 19.

Many colleagues have been involved over a number of years in development of the simulation models described in this book. Those making significant contributions are mentioned below.

Dr Felix Favorite and Ms Patricia Livingston have contributed greatly, especially in evaluation, validation, and presentation of the results and in the operation of the computerized model. Ms Majorie Gregory has typed this book and numerous other related technical reports and Ms Carol E Hastings drafted the figures.

Special thanks are due to other professional colleagues outside the laboratory, some of whom have spent many weeks with the authors in reviewing this work and guiding it with advice, as well as reviewing the present manuscript. Among these colleagues, the following deserve special thanks: Drs K P Andersen and E Ursin of the Danish Institute for Marine and Fisheries Research, Dr N Daan of the Netherlands Institute for Fisheries Research, and Dr E Henderson of the NMFS Northeast Fisheries Center.

The manuscript of this book was thoroughly reviewed by Professor John Magnusson, University of Wisconsin, Madison, Dr John Radovich, California Department of Fish and Game, Sacramento, and Dr Richard Marasco, NWAFC, Seattle, who made numerous suggestions for its improvement and whose invaluable help is herewith acknowledged.

2

Ecosystem processes and approaches to quantitative simulation of the marine fish component

In this chapter we give introductory explanations on four subjects pertaining to marine fisheries ecosystem and its simulations. First, we define briefly the marine ecosystem, list the essential processes in it, and discuss briefly its instability. Secondly, a brief history of development of quantitative evaluation of this system is given. Thirdly, shortcomings of the single-species fish population dynamics approaches and their application to fish resource evaluation are discussed with examples from developments in the North Sea. Fourthly, the objectives and principles of our numerical ecosystem simulation are listed and some definitions pertaining to this simulation are given.

The marine ecosystem, which is largely hidden from the human eye, is complex with respect to its species composition as well as to processes occurring within it. It consists of varying numbers of different species of plants and animals (the biota); processes altering the distribution, abundance, and other properties of this biota; and, the physical and chemical environment which changes in space and time. The apex predators (mammals, birds, and man) are an important part of this ecosystem.

Although the main consideration in this book is the marine fish ecosystem, we cannot fully separate it from the rest of the marine ecosystem. We must consider the production of organic matter by phytoplankton, which affects to a large extent the general productivity or 'carrying capacity' of a given region. The zooplankton serves as food to many fishes, especially to juveniles of most fishes. Another food resource is the animals on and in the bottom – the benthos. Marine mammals, birds, and man are the 'apex predators' of the fish ecosystem, removing fish by predation from given areas and returning parts of the metabolic products and 'wastes' as nutrients in other areas.

There are numerous and diverse processes at work in the marine ecosystem, affecting its biological components in a variety of ways. The major processes affecting the biomass of a given species are: (*a*) growth, which is influenced by temperature and availability of proper food; and changes with age of the species; (*b*) recruitment, which is dependent on spawning success, and is a complex process depending on the size of spawning stock and the mortality of eggs and larvae from various causes, including predation; (*c*) post-larval and juvenile fish predation by other species and by cannibalism; (*d*) mortality from old age, spawning stress, and disease; (*e*) migrations, including immigrations and emigrations in respect to a given region; (*f*) feeding, including food requirements for maintenance and growth and space-time variation of food composition; and, (*g*) apex predation by birds, mammals, and the fishery. One process which controls the ecosystem to a large extent is predation, *ie* one species uses another as food. That big fish eat small fish was well known centuries ago; however, the influence of this process was not fully appreciated in the 'single-species' approach.

The marine ecosystem is not stable: considerable fluctuations occur in abundance and distribution of many species. One of the main tasks of fishery scientists is to determine the abundance of species; their fluctuations and changing distributions; and the factors controlling them. In some ways, however, the complex processes controlling the abundance of species in the marine ecosystem run a relatively steady course over long periods of time. For example, it is quite remarkable that from about 200 000 eggs, released during spawning by a female walleye pollock, in a long time average only two fish (a male and a female) survive to 'mean spawning age', say to age four. However, deviations from this remarkably constant process of reducing the numbers of survivors occur and produce variations in year-class strength of any given species which can be a few to a few tens of times higher or lower than the average. A consequence of a stronger adult year-class, of *eg* pollock or cod, would be that they would eat more herring; thus, the herring population might decrease as a result of increased predation pressure from pollock or cod. The fluctuations in the abundance of species can be caused by numerous other factors in the ecosystem, such as environmental anomalies and/or factors inherent in the populations themselves (*eg* cannibalism).

Although populations of some species may decrease while others increase with time, the standing stock of the total biomass of finfish in a given region fluctuates relatively little. The total biomass is determined by the total availability of phytoplankton, zooplankton, and benthos as the bulk food and determine the so-called 'total carrying capacity' of any given region.

Obviously the fishery will also cause changes in the abundance of fish. Not only the target species will be affected, but also other species caught incidentally or fish interacting as prey, predator, or competitor with the target species.

To form a quantitative picture of changes and interactions in the marine ecosystem, we should assemble available applicable knowledge into simulation models, with the result that analyses of vast amounts of data can be handled only on large computers (with core sizes of 100 000 words and larger). The quantitative computation of changes in the ecosystem requires the use of numerous explicit equations, each adapted to reproduce quantitatively a given process. Generalized flow diagrams of ecosystem simulations permit illustration of some of the emphasis in and peculiarities of the simulation approach. *Figure 1* presents the principal processes which are emphasized in simulation models

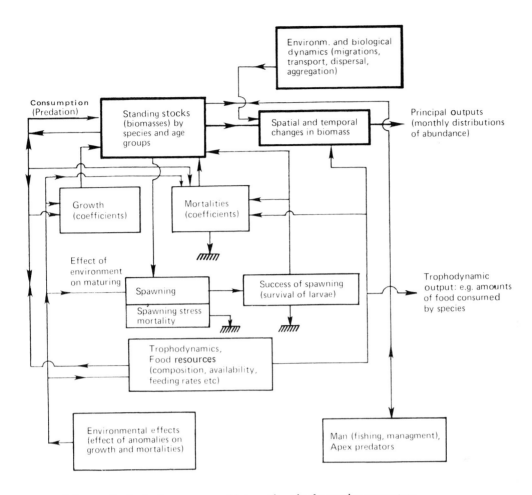

Fig 1 Scheme of principal processes and interactions in the marine ecosystem.

developed at the NWAFC. The heavily lined boxes indicate the main conditions and processes which the ecosystem simulation attempts to solve quantitatively. The other boxes represent processes and interactions in the ecosystem which are directly or indirectly determinant to the main conditions. The arrows present the flow of influence. Arrows ending without terminal boxes indicate that some removal of biomass from the ecosystem can occur.

Pioneering attempts to develop mathematical models for fish and other animal populations, based on predator-prey relations, were conducted prior to 1925 by Alm, Baranov, Ross, and Kevdin. Simple quantitative plankton production models, where different 'trophic levels' in the ecosystem were connected via food requirements, were developed in the 1940s. The multispecies theory of fishing was initiated in the 1950s in connection with the further development of single-species population dynamics models

(Beverton and Holt, 1957). More complex ecosystem simulation was impossible at that time because the large computers required for this task were not available. Numerous minor changes, improvements, and additions have been suggested later in these single-species models, some of which have found application as auxiliary methods, such as cohort or virtual population analysis. As computers were not generally available to fisheries scientists in the 1950s, Beverton and Holt had to weigh the importance of factors on what to include and what to leave out of these formulations. The detailed consideration of predation was included in one parameter – M – the 'natural mortality', partly because a detailed computation of predation was a formidable task for manual computation and partly because the space and time variable composition of food of the species was not well known at this time.

The Beverton-Holt (1957) single-species assessment model is based on three basic formulas (although many more auxiliary formulas are presented):

$$dN/dt = (F + M)N(t) \qquad \text{(change of numbers)} \qquad (1)$$

$$dw/dt = Hw(t)^{2/3} - k_m w(t) \qquad \text{(growth)} \qquad (2)$$

$$dY/dt = FN(t)w(t) \qquad \text{(yield)} \qquad (3)$$

where t is time, N–number of fish, w–body weight, Y–yield, F–fishing mortality, M–natural mortality, H–food requirement for growth parameter, and k_m–food requirement for metabolism parameter.

In single-species models, each species is treated separately as if other species do not exist; thus one species cannot influence the abundance of the others by predation. Another shortcoming in the application of the above formulation is the lack of data for determination of M–the natural mortality. As we know now the natural mortality is largely a function of age and size of the fish; the greatest part of M in juveniles is predation mortality, and in adults it is the spawning stress and 'old age and disease' mortality. Single-species models do not consider distribution. In addition, the various effects of migrations or the effects of changing environmental influences cannot be treated in these single-species considerations.

The intensively fished North Sea provides an example of the ineffectiveness of the single-species model. The roundfish landings increased in the 1960s, but surprisingly the biomass (stock) also increased (*Fig 2*). No single-species model did or could predict this biomass increase because when the fishery increases, the stock decreases according to these models.

On the other hand, the stocks of mackerel and herring decreased at the same time (*Fig 3*) due to the intensive Norwegian fishery on these species. Thus, it appeared to Ursin (1979) that mackerel and herring might have been preying on roundfish larvae, and when the stocks of mackerel and herring decreased, due to the intensive fishery, more roundfish larvae survived.

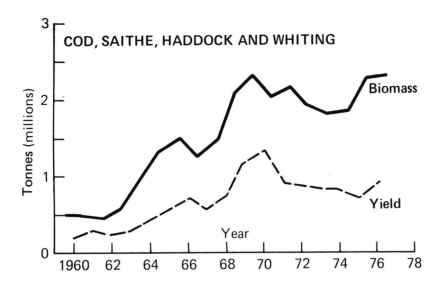

Fig 2 Biomass and yield of cod, saithe, haddock, and whiting in the North Sea from 1960 to 1976 (Ursin, 1979).

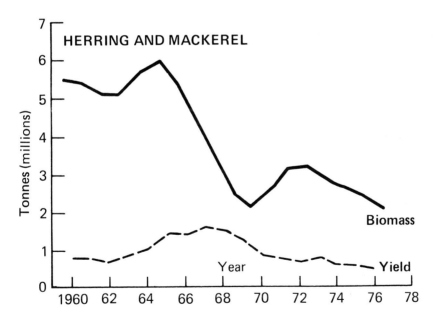

Fig 3 Biomass and yield of herring and mackerel in the North Sea from 1960 to 1976 (Ursin, 1979).

Andersen and Ursin (1977) developed an extensive multispecies model applicable to modern fisheries management problems by adding one more basic term to Beverton and Holt's (1957) basic formulas:

$$dR/dt = f(t)hw(t)^{2/3} \qquad (4)$$

where R is food consumed, f is a variable feeding level, and h is a coefficient. With this model they were able to show, for example, that if the fishing effort on cod in the North Sea was reduced to 30% of the existing (1977) level (*ie* to achieve maximum yield with minimum effort according to a single-species model – *Fig 4* (Ursin 1979)), the biomass of cod would increase fourfold (not sixfold as predicted by the single-species model), but without any increase in total yield because the stocks of medium-sized species, which are commercial food fish such as haddock, would shrink to half of their present stock size mainly due to predation by cod.

The above-described behaviour of the fish stocks in the North Sea shows that they

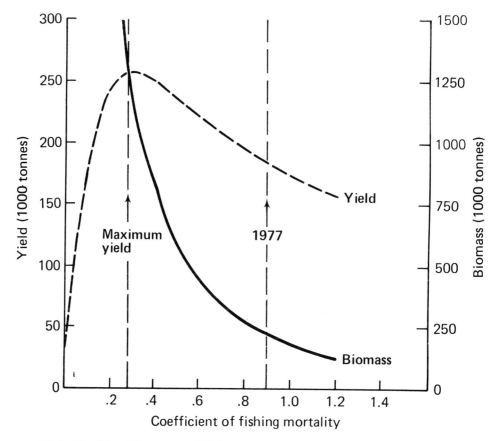

Fig 4 Traditional Beverton and Holt assessment for North Sea cod (Ursin, 1979).

10

have not followed conventional fisheries population dynamics forecasts. In addition, many pelagic fish stocks in different regions of the oceans have collapsed rather suddenly to very low levels in the last three decades and, in many cases, the single-species considerations did not fully indicate the imminence of such collapses. Most fishery biologists are now convinced that the marine fish ecosystem is rather unstable and that 'natural fluctuations' of stocks occur, some of which can have considerable magnitudes and long periods. The mechanism and 'speed' of stock recovery is also poorly known and no useful quantitative information about recovery speeds can be obtained from single-species models. Thus, we can conclude that there is a need for a comprehensive method for evaluation of the dynamics of the marine fish resources. We have attempted a possible solution to this problem by summarizing quantitatively our existing knowledge of the marine fish ecosystem and simulating it quantitatively in a dynamic numerical model.

The multispecies model of Andersen and Ursin (1977) is the first extensive simulation model which seeks numerical solutions for established formulations. Various forms of their simulation use 14 to 81 entities of plants, animals, and nutritive matter, calling for simultaneous solution of from 42 to 308 differential equations. The model emphasizes trophodynamics; growth rates of all species are a function of the season and availability of food; and natural mortality has been partitioned into various components such as predation mortality, spawning strain, starvation and disease mortalities.

The first multipurpose ecosystem model for large estuaries (for water quality management) is the GEMBASE model (General Ecosystem Model for Bristol Channel and Severn Estuary – Longhurst, 1978; Longhurst and Radford, 1975), which although initially formulated for water quality management was also designed as a tool for ecological studies. This model simulates the carbon and nitrogen flow among ecological state variables and seven geographical regions. The whole process requires about 150 equations with 225 parameter values. It uses hydrodynamical models for transfer of materials between adjacent geographical regions but does not deal fully with fish components.

Several numerical two- and three-dimensional ecosystem models have been developed recently which deal essentially with planktonic organisms as the basis for marine productivity (eg Kremer and Nixon, 1977). The nutrient-plankton-fish energy pathways are, however, greatly variable in space and time, with great lateral losses (eg losses into deep water, remineralization, etc) that are not yet fully accounted for quantitatively and hamper the use and interpretation of plankton production based models. Such models require extensive data bases, that are not available in many regions where real or potential fishery problems exist, and are generally applicable to small regions.

The most pressing present-day fisheries research problem in the NE Pacific confronting the NWAFC (Northwest and Alaska Fisheries Center in Seattle) is the assessment of the status of the fish stocks and their fluctuations. The results of such studies are required for use in fisheries management of the '200-mile zone', including resource allocations to foreign countries. The NE Pacific encompasses numerous productive fishing areas throughout the eastern Bering Sea, Gulf of Alaska, and west coast of the United States – an area of about 2 million square kilometers. Fisheries development in this region has been relatively recent (starting at the end of the 1950s) and fisheries research in this vast area has been limited. In recent years regular trawling surveys have been carried out in the eastern Bering Sea during the summer months and more occasionally in the Gulf of Alaska and along the west coast of North America. These costly surveys can give only a general picture of the abundance and distribution of species vulnerable to catch with bottom trawl. Grosslein (1976) has shown that, in general, the accuracy of estimates based on trawling survey results is at best $\pm 50\%$.

Although single-species stock assessment models have been applied to NE Pacific commercial species, the results of such applications are less reliable than in other well explored and exploited areas, such as the North Atlantic. There are several reasons for this. For example, the Virtual Population Analysis (VPA) is not fully applicable in the NE Pacific because: (a) the catch statistics and length-age frequency distribution data are haphazardly and incompletely collected by a number of nations involved in the fishery; (b) most species undertake extensive seasonal and 'life cycle' migrations; (c) the natural mortality rate estimates are very unreliable; (d) most species are under-exploited; and (è) there is a heavy consumption of fish by marine mammals (they consume twice as many fish in the eastern Bering Sea as the total commercial catch).

Similarly, any other single-species model is not fully applicable to the stocks in the NE Pacific because: the initial stock size is not known, year-class strength determination is deficient due to insufficient data, age (size) specific natural mortality is not known, and fisheries statistics are deficient. An analysis of shortcomings of past resource assessment models (including a recent analysis of these problems by Dickie, 1979) and the review of types, availability, and reliability of basic data of NE Pacific fish and fisheries, suggested that a synthetic and holistic ecosystem approach for resource evaluation should be undertaken in which use is made of most available data and knowledge.

We decided to develop a biomass-based simulation model for marine ecosystem simulation as it presented to us some advantages over number-based models (see next chapter). The initial NWAFC ecosystem simulation models, the Prognostic Bulk Biomass (PROBUB) and Dynamical Numerical Marine Ecosystem (DYNUMES) models (Laevastu and Favorite, 1978a and 1978b), were developed independently without any prior knowledge of the Andersen-Ursin model. While there are some differences and some similarities in approaches in these models, the results should be comparable if applied to the same problems and with the same reliability of input data. Whereas the Andersen-Ursin model obtains the initial input of year-class strength in numbers from single-species considerations and from cohort analysis and computes to a steady-state solution with this given input, the NWAFC's PROBUB model searches for a unique solution with a given set of inputs, ie determines the abundance of various species biomasses in a defined equilibrium state. In a prognostic mode, the PROBUB model allows the determination of various fluctuations in abundance of species caused by environmental anomalies as well as by the fishery. The DYNUMES model has spatial resolution which allows the simulations of seasonal migrations and their effect and spatial and temporal variation of composition of food.

Every large-scale system simulation requires the establishment of objectives and basic principles while the system is conceptually designed. The overall objective of ecosystem simulation is to reproduce quantitatively the essential conditions and processes in the marine ecosystem in a time-dependent mode.

The objectives of present numerical ecosystem simulations at NWAFC, which guided the simulation development, can be grouped into two main categories described below: investigative and digestive, and general management guidance.

Investigative and digestive (analytical) objectives include basic ecological research that permits quantitative determination of the state of the ecosystem, the evaluation of available knowledge, and establishment of priorities for future research. These objectives establish that the simulation model be designed to synthesize available pertinent knowledge of the ecosystem in a reviewable manner. Furthermore, the simulation should serve as a research tool in ecological research, especially for the study and quantitative determination of interspecies interactions and the determination of the effect of environment and its anomalies on the fishery resources.

General management guidance objectives include the assessment of fisheries resources, and the determination of the effects of exploitation on these resources. These objectives require that the simulation must be able to determine the magnitudes and present status of the marine living resources in a defined region. Furthermore, the simulation would be used for quantitative evaluation of the response of the resources (and total ecosystem) to the space and time varying fishery, either present or projected; thus the simulation must be time-dependent.

There are some basic requirements of ecosystem simulation, the fulfillment of which is essential for realistic simulation:

——The simulation must include all components of the biota. This is necessary for realistic simulation of trophodynamic processes (feeding) and processes dependent on feeding (such as growth). Due to computer core and time limitations, several species are often combined into groups of ecologically similar species. The simulation must also include all essential processes in the ecosystem, including environmental processes which affect the biota.

——Simulation must have a diagnostic phase (*ie* analysis of initial conditions) and a prognostic phase with proper time steps.

——It must have a proper space and time resolution, which is defined by the region under consideration and by available computer resources.

——Mathematical formulas in the simulation must serve for quantitative reproduction of known distributions and processes (*ie* to simulate the known and proven). This means that the simulation must be deterministic and based on available data and accepted formulas. Theoretical conceptualization, so common in modeling, should be avoided unless the theory has been tested and proven by empirical research.

——Explicit approaches, free from mathematical artifacts, must be preferred, (*ie* the mathematical formulas used in the model must reproduce known processes, consistent with data and functionally logical, rather than assume that a mathematical formula represents the behaviour of a system).

——Biomass balance and trophodynamic computations must start with apex predators (including man); these can be treated as 'forcing functions' of the system.

In addition, the following requirements apply to ecosystem simulations which emphasize 'fisheries ecosystem':

——Simulation must be capable of solving the major part of the age-variable mortalities (especially predation, spawning stress, and fishing mortalities).

——There must be a unique solution to the system of basic equations in defined conditions (for determination of the 'equilibrium biomasses', *ie* the plausible magnitudes of the resources).

——The system of equations should not be conditionally stable (except for unique solutions in defined conditions), as a marine ecosystem is not stable but fluctuates within varying limits.

——Singular components of the system (state variables, processes, rates, *etc*) must be verifiable at any time step and location.

——Simulation must include migrations by various causes as well as random movements. These migrations will affect the biomasses by spatial and temporal changes of predator-prey relationships.

——Total carrying capacity must reflect the space-time variable plankton and benthos production (the 'production buffers').

——Major environmental factors, such as temperature, currents, *etc* must be included in the simulations and reflect the prevailing knowledge of their space-time variations.

——Outputs of the simulation should be available for all parameters in desired space-time resolution.

——The simulation must be tailored to the availability of local data and local knowledge.

The ecosystem simulation levies the following requirements on the simulation development and use team:

——Establishment of goal and problems to be solved.

——Thorough and wide knowledge of all species, ecology, and environment (a team of 'all around', goal-oriented scientists is required).

——Knowledge of applied mathematics and its limitations.

——Experience with large-scale computerized models.

————Use of larger computers (minimum core size 100 000 words).

————Knowledge and access to all regional and general data required in the simulation.

The principles of trophodynamic computations are schematically shown on *Fig 5* where the arrows show the flow of food and dots, squares and triangles indicate the origin of food. These computations permit determination of who eats what and how much and, consequently, how much of each (consumed) species must be there to produce the eaten amounts, using empirically determined growth coefficients.

In this approach the amounts consumed must be determined in a previous time-step (a time-step earlier than computation of growth), rather than in a time-centered (simultaneous) fashion. Thus, the model must be based on a finite-difference approach with a combination of backward and forward time-stepping. Such numerical techniques are used in meteorological modeling. The length of the time-step must be evaluated with respect to the errors (both in time-stepping technique and in the data) caused by the length of the time-step. We found that the maximum time-step to be used is one month; some approaches in the model do require shorter time-steps due to numerical stability requirements.

For any geography related simulation, the applicable space must be defined. The simplest space definition is by geographical regions ('boxes'), such as fisheries statistical

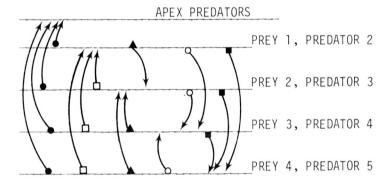

APEX PREDATORS

PREY 1, PREDATOR 2

PREY 2, PREDATOR 3

PREY 3, PREDATOR 4

PREY 4, PREDATOR 5

Principles: Determine who eats what and how much and then determine how much of the prey must be there to produce the eaten amounts.

Advantages: Minimum values of the production and standing stocks of all prey can be computed.

Amounts of noncommercial (and nonsampled) species can be estimated.

Changes in one prey biomass are related to changes in other prey biomasses.

Fig 5 Principles of trophodynamic computations, based on consumption.

areas (*Fig 6a*). In such 'box models' without further space resolution than geographic regions, the defined areas are assumed to be homogeneous; a distinction is made between shallow areas (< 200m depth) and deep areas (> 200m depth). The outer boundaries (in most cases 200 nautical miles from the coast) are rather arbitrary in respect to the biota. Consequently, all quantitative computations must be normalized on unit area (*eg* km²). All seasonal migrations, which could occur through the boundaries of the regions, must be estimated and prescribed. The errors arising from such estimates can be minimized by considering large, relatively confined regions as units such as the Bering Sea and northern Gulf of Alaska.

The numerical simulation of an ecosystem in three-dimensional space requires two-dimensional grids with defined grid size (see example of DYNUMES grid with 97·5km grid size in *Fig 6b*). In these gridded models all computations are carried out at each grid point (grid intersection) and time-step with prevailing conditions at each grid point location. The advection and migrations occur from grid point to grid point in u and v (*ie* right–left and up–down) components. This grid can be repeated for several depth levels (*eg* near–surface layer and bottom), presenting therewith depth as third space dimension. All space-dependent input data are digitized at each grid point and outputs are given for each grid point as well as in numerical form.

The term ecosystem simulation used in this chapter should be defined for clarity. In accordance with Henri Poincaré: 'Clarity in argument can be obtained if it has been introduced into the definition'. Our definition of ecosystem simulation is: '**Quantitative (numerical) reproduction of a marine ecosystem and its dynamics, based on the best available quantitative knowledge of this system.**' Furthermore, holistic ecosystem simulation has been used as synonymous to ecosystem simulation. The holistic (structural, deterministic) simulation is defined as: '**Numerical quantitative reproduction of a system by structural parts of it, using deterministic formulations justified by empirical data.**' The latter definition fits our approach as well, although the first definition is a more general one.

Thus, in contrast to mathematical (and/or statistical) modeling, which often involves assumptions that a process and/or system behaves as a given mathematical formula, our simulation is based on empirically derived and tested formulations which are logically consistent with known facts and which can be singular models *per se*. Further, in an ecosystem simulation the different species and environments are permitted to interact with each other in various known ways which produce results not found and/or considered before; often because their derivation would have been too laborious and/or hidden by various interactions.

Differentiation by degree of complexity between ecosystem models and ecosystem simulations is useful although these terms have been used as synonyms. We usually consider a model to be an abstraction and simplification of a condition or process, whereas a simulation is a reproduction of a system of conditions and processes based on available empirical data and may contain many models.

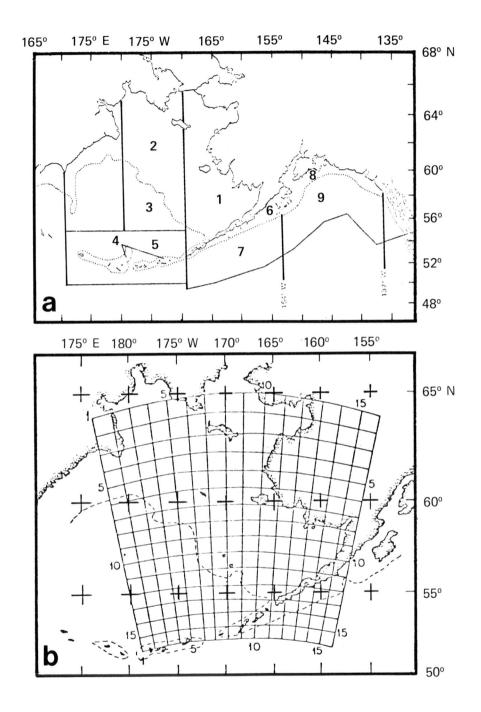

*Fig 6*a – Fisheries management and statistical areas used in PROBUB model and b–Computation grid of DYNUMES model for the eastern Bering Sea.

3

Limitation of models based on primary production, and comparison of biomass-based and number-based models

In marine ecology it has been customary to consider the marine ecosystem in a given sequence of the food chain (also called food pyramid), starting with the production of the organic matter (primary production) by plants (mainly phytoplankton), following it through the food chain which ends with the highest order of predators (so-called apex predators). In this chapter we examine the processes of primary production and possibilities and shortcomings of their quantitative simulation. Thereafter we describe some advantages of starting the marine ecosystem simulation with the apex predation and working stepwise towards the basic organic production – *ie* we use an 'upside down' ecosystem modeling approach.

In the second part of this chapter we examine the advantages of a biomass-based ecosystem simulation as compared to a number-based simulation, and describe some of the peculiarities of biomass-based simulation.

The term production has been widely used in connection with the dynamics of organic matter in the marine ecosystem. The term normally implies dynamics, thus one must define the subject of production and quantify production as a rate, *ie* define the unit of time, amount, and entity considered. The most common use of the term production is in the production of organic matter by plants, mainly phytoplankton ('basic organic production').

Organic production depends first on the various types of producers and their production capacity. Thus, to estimate production we must know the type (species) and the quantity present at the initial time or at any given time step, *ie* we must know the standing stock (or standing crop) of all the producing species and the rate of production of each species and/or groups of species. The standing stock is changing with time and depends on the reproduction of itself (thus a nonlinear second order interaction exists). Furthermore, the producers are being eaten and die from other causes (these processes also have time and space scales), creating secondary nonlinear effects.

In addition to the age of the individuals in the population, production capacity depends also on the following conditions:

    ——Availability of matter (nutrients) required to produce organic matter (*ie* carbon, oxygen, and other minor constituents). Thus the quantities of these substances must be measured at given locations and times if one is to compute production in space and time. Further, the availability of these quantities is dynamic, depending on transport and regeneration of the produced organic matter. The required density of measurements in space and time for use in reliable simulation are not available for these quantities and processes for any region of the ocean.

——Energy is required for production of organic matter. Basic organic production requires light energy and this is dependent on the energy available on the surface (which in addition to being dependent on location and time is also dependent on variations in cloud cover). The availability of light energy at depth depends on the depth and location as well as the turbidity of the water and its change with time. Numerous other complicating factors occur: the dynamics of the producers and its consequences (*eg* aggregation, patchiness, *etc*) markedly influences the turbidity, different organisms utilize different wave lengths of light, and the energy utilization (or the production capability) depends also on temperature. The space and time variation of light energy can only be approximated at present (*re* rapid change in cloudiness, *etc*).

Most of the parameters necessary for the computation of basic organic production can be estimated at any given location and time to only an order of magnitude. These estimates do not warrant a complex formulation of the process and do not allow that the result of this process (*ie* the amount of organic matter produced) would be determinant for other (following) processes (*ie* utilization of organic matter by animals). Furthermore, the pathways of the produced organic matter in the marine ecosystem are greatly variable in space and time, *ie* the utilization and regeneration of organic matter is highly variable. Thus, one could expect only approximate results if quantitative ecosystem computations are based on the estimation of basic organic production and its utilization.

A schematic presentation of a conventional food web (or food pyramid) model is given in *Fig 7*. In addition to the above described limitations on the use of such simple models, it should be added that no well-defined trophic levels exist in the marine ecosystem, rather oceanic food webs are quite intermixed (Isaacs, 1976). Furthermore, these simple food web models preclude the biomass of secondary consumers from exceeding the biomass of herbivores in a steady state. In fact, however, there are larger biomasses of secondary consumers than there are primary consumers in the marine ecosystem (Isaacs, 1976).

At present we can use organic production measurements and estimates only for comparison of different regions as to their 'productivity' in large space and time scales. Some researchers have attempted to evaluate production in the marine ecosystem in terms of energy and 'energy flow' instead of biomass. This procedure only adds considerable uncertainty to conversion factors and makes the comparison with reality a virtual impossibility.

We can often find in the literature the term 'secondary production' which usually

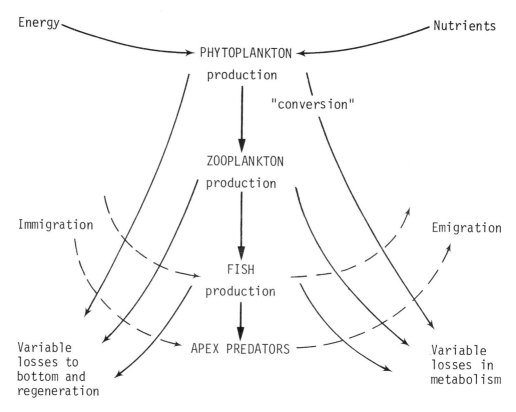

Energy

Nutrients

PHYTOPLANKTON

production

"conversion"

ZOOPLANKTON

production

Immigration

Emigration

FISH

production

Variable
losses to
bottom and
regeneration

APEX PREDATORS

Variable
losses in
metabolism

Principles: Estimation of organic production and its "conversion" to
              other trophic levels
Shortcomings: Primary production sampling incomplete.

              Conversion efficiencies variable and badly known

              Magnitudes of "losses" nearly unknown.

              Proportioning between trophic levels highly variable.

Fig 7   Principles of a conventional food chain or food pyramid model based on production.

refers to the dynamics of the biomass of zooplankton and sometimes even fish. Even a less accurately definable term, 'surplus production', is found in the fisheries literature. Examination of biomass dynamics has led to the conclusion that we can consider biomass conversion from one form or species to another rather than its 'production'. In the process of biomass conversion we can recognize production utilization, utilization efficiency, and/or conversion efficiency as distinct parameters.

The first step in the utilization of any biomass in the sea is the process of consumption, live or dead. The second process is the digestion of the matter which results in the conversion of it into new biomass, breakdown to basic nutrients and organic compounds (non living), and release of some energy for activity as well as other living processes. Some eaten matter is excreted undigested which can be used by other organisms, including bacteria, in the ecosystem.

The conversion of the digested matter into a new form of biomass thus results in growth via catabolic processes and the breakdown of organic matter (anabolic processes), which together are called metabolic processes. The utilization of organic matter for energy release is also a process that regenerates basic nutrients. Thus, from the point of view of dynamics of organic matter, we must consider the processes of consumption (including predation), growth, reutilization (disposal) of growth (including sex products development and release), and remineralization.

Predation (dealt with more fully in another chapter) depends on many conditions of the prey and predator as well as general ecosystem conditions. Among these determining factors are: suitability (preference) of prey to predator, predator-prey size relations, availability of prey to predator (encounter-density dependent), and avoidance behaviour of prey. Thus, mobility and distribution of predator and prey in space and time are important factors determining predation.

Growth of individual species and of population biomass is a complex process affected by many factors and is described in detail in another chapter. First, it is dependent on the efficiency of digestion (ie how much of the eaten material is converted to tissue), thus dependent on the physiology of the organism and on some physical factors, such as temperature. As a population consists of individuals of different ages and sizes, the growth rate of any individual is different from the average growth rate of a total biomass. Obviously the amounts of food available and taken are additional important biomass growth determinants. Finally, we must consider the disposal (utilization) of growth of any biomass and arrive by necessity to an 'inward' decreasing circulation system of total biomass in the marine ecosystem which is schematically shown in *Fig 8*. The circulation of the organic matter does not end with the death of an organism – the carcasses are utilized to a large extent by many groups of animals, especially demersal fish and benthos.

We can conclude from the summary discussion of the circulation of organic matter (biomasses) in the marine ecosystem that there is no definable surplus production in the marine ecosystem except very minor amounts buried in the sediments. Furthermore, there are no clearly defined food chains and trophic levels either – this concept has been grossly oversimplified in the past. Thus, we can also conclude that no single or simple theory, and no simple primitive equation formulations can reproduce the marine ecosystem. This system must be simulated with a set of equations which reproduce individual processes and distributions. However, three important processes dominate quantitatively in the marine ecosystem, controlling to a considerable extent the abundance and distribution of individual components: growth, predation, and migration. Each of these processes is, in turn, controlled by a relatively complex set of conditions.

The above conclusions are demonstrated below in mathematical notation with simplified assumptions. The biomass of any given population is a function of time and location P=f (t,x,y,z). The individual time change of this population (dP/dt) is:

$$\frac{dP}{dt} = \frac{\partial P}{\partial t} + \frac{\partial P}{\partial x} V_x + \frac{\partial P}{\partial y} V_y + \frac{\partial P}{\partial z} V_z \qquad (5)$$

The first term on the right side ($\partial P/\partial t$) is the local time change and the following three terms represent migrations and advection. These migration terms comprise several processes, such as spawning and feeding migrations, dispersal and aggregation, migrations caused by unfavorable environmental conditions, and transport by currents. The migration terms determine largely the spatial distribution of most species/ ecological groups. These terms never vanish; however, if we consider a large region (such as the Bering Sea), we may assume that the migrations into and out of this large region are negligible and that the migration term is 0. In this case, individual change equals local time change. The local time change is largely a function of biomass grown (G) and its removal (Q), the latter comprising mainly predation, mortality and, the fishery.

$$\frac{\partial P}{\partial t} = f(G, Q) \qquad (6)$$

If we assume that an equilibrium might exist (which, however, is very unlikely) in an

ecosystem over one year, then $\frac{\partial P}{\partial t} = 0$ and $dG = dQ$ \qquad (7)

Formula 7 is the criterion for finding 'equilibrium biomasses' with the deterministic PROBUB model (see Chapter 8).

Biomass growth of any given species is a function of the age of the species (A_i), time of the year (t) (also in relation to *eg* spawning), food availability (Q_f), and environmental conditions (T).

$$\frac{\partial G}{\partial t} = f(A_i, t, Q_f, T) \qquad (8)$$

The removal of biomass is a complex space-time function of many variables, such as food requirements of many components (species) of the ecosystem ($q_i \ldots q_n$), mortalities from various causes (Z...), *etc.*

$$\frac{\partial Q}{\partial t} = f(t, x, y, z, q_1 \ldots q_n, Z \ldots) \qquad (9)$$

Thus, it becomes apparent that quantitative computation of changes in the ecosystem requires the use of many explicit equations. Consequently, the numerical ecosystem simulation becomes, to a large extent, the accounting of growth, removal (predation and mortalities), and the change of biomasses in space and time.

The last two generations of marine and fisheries scientists have attempted to evaluate quantitatively the standing stocks and production in the sea. One of these attempts (by Laevastu), prepared in 1956, is shown in *Fig 9*. This evaluation attempt is based on the assumption that benthos and zooplankton production (double circles) are known and food requirements and average composition of food are also assumed to be known. Such schemes indicate the complexity of marine food chains and can give only an approximate, idea of actual production of fish. Realistic and useful quantitative fish ecosystem simulation can be accomplished now by computerized simulations as described in this book.

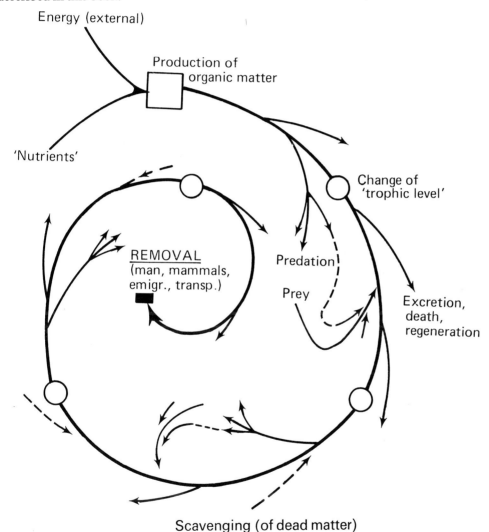

Fig 8   The 'biomass spiral'.

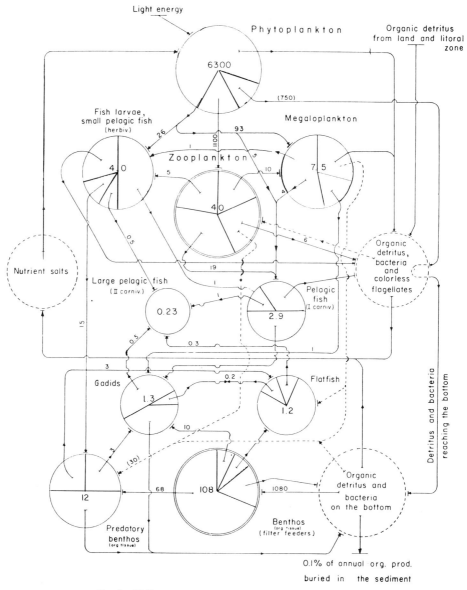

Assumptions : Depth 100 m.
Standing crop of zooplankton 200 mg/m³;turnover rate,(annual) 2
Standing crop of benthos 100 g/m² ;turnover rate,(annual) 1.5

The number in the circle indicates the annual production of biomass in g/m²
The number on the lines indicates the amount of biomass in g annually consumed by corresponding groups.

Fig 9 An example of a balanced food chain in the sea, as computed with simplified assumptions.

24

Traditionally, fisheries population dynamics studies using single-species models have been number-based, whereby the population has been divided into cohorts (year classes). The numerical strength of each year-class has been computed separately, thus requiring separate book-keeping of each year-class. In longer time series computations, representing more than one year in real time, a semi-discontinuity occurs at the time of spawning, where the first year-class becomes the second, the second becomes the third, *etc.* Some difficulties (and inaccuracies) are encountered in number-based models in simulating the spawning, as in most species actual spawning will last several months.

As the growth rate of a given species can vary considerably in space and time and as spawning can last several months, a given year-class is represented by fish of various size, and some inaccuracies are introduced in number-based models in computing growth rate changes and feeding (predation), which is greatly size dependent. These difficulties are magnified in the computation of the 'number reduction' (*ie* predation, other mortalities, and fishery), especially in juveniles, from one year-class to another, and particularly when a smaller time step than one year is used. Although some of the above difficulties can be alleviated by dividing a given species population into age groups of less than a year, this would be impossible in large ecosystem simulation, due to computer core limitations, as all species in the ecosystem must be included. Undoubtedly the use of number-based models will continue in fisheries population work, especially in single-species considerations, despite the above mentioned difficulties.

On the other hand, biomass-based models have been seldom used in the past in fisheries population studies, although they have been used in various 'food chain' models. Biomass-based models offer some advantages, as compared to number-based models, especially in trophodynamic computations and in fish ecosystem simulation in regions where conventional fisheries data are deficient. It is not necessary to divide a biomass of a given population into age classes in biomass-based models, as the age class strengths can be reconstructed with sufficient detail at any given time step in properly designed ecosystem simulations. This reconstruction requires that the growth rate of biomass and fraction of exploitable biomass are properly accounted for. The recruitment in space and time can also be simulated with considerable complexity in biomass-based models (see Chapter 5 Section 5.5). Furthermore, possible uncertainties in the simulation of recruitment in biomass-based models are 'buffered' by the presence of the biomass of the previous year-classes.

Some of the differences between number- and biomass-based considerations are partly apparent from *Fig 10* which shows the distribution of numbers and biomass with age in Pacific cod from the Gulf of Alaska. The numbers and biomasses have been normalized to the interval of 0·5 to 9 years.

The distribution of numbers with age in juvenile (prefishery) age-classes must be well known in number-based considerations in order to achieve desired accuracy in exploitable age groups in time dependent computations. Therefore, in number-based models it is necessary to divide the stock into age groups and operate with discontinu-

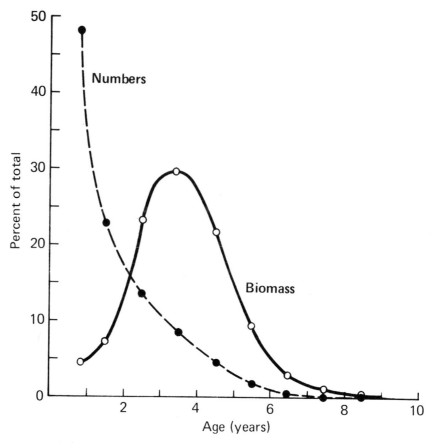

Fig 10 Distribution of numbers and biomass with age in Pacific cod in Gulf of Alaska (from 0·5 to 9 years).

ities, when age group time limits have been exceeded (*eg* when two-year-olds become three-year-olds, *etc*). This is not necessary in biomass-based considerations, partly due to the fact that the maximum of the biomass distribution is usually at an age when fish come under exploitation. Furthermore, only one growth rate for the whole biomass needs to be used in various computations in the ecosystem simulation with biomass-based models. The change of the biomass distribution with age in the case of stronger and/or weaker year-classes is described in Section 5.5 Chapter 5 (*eg Fig 18*).

The growth rate of biomass is one of the important space-time variable parameters in biomass-based models. Its changes due to environmental influences (*eg* temperature) and/or partial starvation can be easily simulated according to available empirical knowledge. The space and time variable food composition can also be simulated more easily in biomass-based models than in number-based models.

Biomass-based models require the use of finite difference methods, usually involving backstepping. Implicit methods are, however, also used in many computations in the simulations.

One of the more important considerations requiring a biomass-based model is the iterative computation of equilibrium biomasses, which would be a formidable, if not impossible, task to compute on number bases as many required parameters are not available for juvenile year-classes.

Biomass-based models can thus circumvent to some extent the difficulties of accounting of numbers of juvenile age-classes in number-based models. The relative (fractional) biomasses of juvenile age-classes are not high in many fishes. Thus, for those species, the errors in juvenile age-class in biomass-based models are partially buffered by the presence of many year classes, as the biomass-based models treat, in most operations, the biomass of a given species as a whole. However, for the purpose of some special studies, it is necessary also in biomass-based models to divide the biomass of a given species into size groups which can correspond to age groups. This is usually done with one species at a time. The recruitment from one size group to another is by size and is thus continuous in time. This is desirable in the light of diversity in growth and also because of long spawning periods of most species in higher latitudes. Experiences with the models where such size-group separations have been made, have shown that the inter-group recruitment is also a rather complex process caused by size (age) dependent mortalities (especially predation) and growth, and requires considerable numerical 'tuning'.

The need for size-group use in simulations is more common in the DYNUMES model than in the PROBUB model because of the necessity of quantitative reproduction of different spatial distributions as well as dynamics of juveniles and adults.

Experiences have also shown that considerable rethinking is necessary in conversion of parameters, concepts, processes, and their results from number-based population considerations to biomass-based considerations.

The biomass-based simulations require some specific biomass parameters, the derivation of which is described in the next chapter.

4

Biomass parameters of fish population and distribution of biomass with age

The ecosystem simulation described in this book is based on biomass rather than on numbers. In this chapter we describe the mean age structure of fish population on the bases of numbers and biomass, and present methods for computation of biomass parameters of fish populations. We also describe a method for computation of total mortality as function of age.

Several concepts of biomass distributions with age and age/size dependent mortality are somewhat different from those used in conventional fish population dynamics. Furthermore, in the following chapter age and size of a given species are used as quasi-synonyms, as size is a direct function of age. In the following chapters the term 'biomass distribution' signifies the quantitative distribution of biomass in different age groups of a given population and/or stock.

Cushing (1976) noted that the age structure of a marine fish population has rarely been described in full. He also pointed out that the trend of natural mortality with age should be known for fish population studies and that a death rate that is constant in age is unlikely because the population cannot be terminated in age. Furthermore, he showed that the age distribution of biomass must reach a maximum at a given age when biomass growth and mortality rates are equal.

Natural mortality of a cohort is expected to change with age (and with growth rates) because predation as a main cause of natural mortality is a predator-prey size dependent process (as has been abundantly demonstrated, *eg* Daan, 1973). There exists also a con-siderable spawning stress mortality for some species (Andersen and Ursin, 1977) or an age dependent senescent mortality (Beverton, 1963; Cushing, 1976), which would increase the mortality rate at older ages. Changes in age specific mortality rates will affect the biomass distribution with respect to age within a population. Furthermore, since growth and mortality rates, and food requirements are functions of fish size/age, the size/age distribution of the whole biomass for a given population must be known to compute these biomass parameters.

When the biomass distribution of a population changes with time, due to the changes in fishing intensity, changing recruitment, *etc*, the corresponding size/age dependent biomass parameters such as growth rate will also change. On the other hand, if some of the biomass distribution with size/age dependent parameters (such as mortality rate) change, the distribution of biomass with age will also change. In addition, the quantitative knowledge of biomass distribution with age permits computation of the fraction of exploitable and juvenile biomasses for a variety of uses (*eg* for fishing/yield computations, for comparison of computed results to survey results, *etc*).

For the computation of the distribution of biomass as a function of age, the following data are needed: weight at age, frequency distribution of exploited age-classes, average numerical relation of adjacent age-classes in the prefishery juveniles (prerecruits), and biomass turnover rate. Empirical data on weight at age are usually available for all species. These data are also used for computation of growth rates. The data on weight at age for both sexes can be combined for these computations. The mean weight of juveniles is often not available and must be extrapolated from data on adults. The von Bertalanffy growth equation, although useful in some earlier population dynamics work, is not useful for the determination of true growth rates because of its inherent inaccuracy in young and old age groups and because it smoothes out some small, but biologically significant, changes in growth rate.

The long-term mean age frequency distribution of fully recruited year-classes is determined from at least 10 years of data to eliminate effects of strong and weak year-classes. Furthermore, the annual means of these data should include data from different seasons and from different fishing grounds in a given region to eliminate bias caused by spatial and temporal variations in age composition of catches in many species. Only the fully exploited year-class numbers should be considered, and those year-classes which are only partially retained by the gear should be eliminated (*ie* 'knife edge recruitment').

To complete the age frequency distribution for prefishery (juvenile) age-classes, an average numerical relation of adjacent age-classes is needed in the form of a factor with which previous older age-class could be multiplied to obtain successively the 'average' numerical strengths of younger age-classes.

Below we attempt to establish an average frequency distribution for marine fish. This distribution would vary from species to species and region to region – depending on growth and mortality rates, fecundity, and life span. However, this average distribution of numbers of juveniles serves only as a first guess in the computations of species- and region-specific number distribution with age. The mean frequency distribution of juveniles is adjusted in computations to produce a biologically plausible distribution based mainly on two conditions:

——The number of fish in a long-term mean age-class must be bigger than the number in the next older long-term mean age-class by the amount of total mortality in the age-class under consideration.

——The mortality of juveniles is mainly due to predation. Vulnerability to predation increases with decreasing size of the fish. Furthermore, the total predation mortality of the species is a function of the turnover rate of its biomass in the ecosystem which, in turn, is largely a function of the growth

29

rate, magnitude of biomass present, and its vulnerability to predation. Support for the concept that predation mortality largely controls the year-class strength is rapidly gaining ground (Rothschild and Forney, 1979).

Cushing (1976) calculated the trend in numbers with age for North Sea plaice (*Fig 11*). His calculation was largely dependent upon the estimate of larval mortality. We obtained additional estimates of number distribution with age by considering the size dependent predation and food requirements, using ecosystem simulation. Using the data on walleye pollock from the PROBUB model for the eastern Bering Sea and normalizing the number of 600 to 1 000g pollock (average age 5·8 years) to two individuals, we found that there were 7·7 fish at an average age of 3·5 years (200g to 600g). We can compute from the food requirement and food composition data that the normalized 3·5 year old pollock population will eat about 2 600 fish of about 4g each (comparable to 6 to 9 month old pollock) and the 5·8 year old normalized pollock population will eat about 25 fish of about 20g (comparable to 1·2 year old pollock). The above normalized numbers are shown with an 'x' in *Fig 11*.

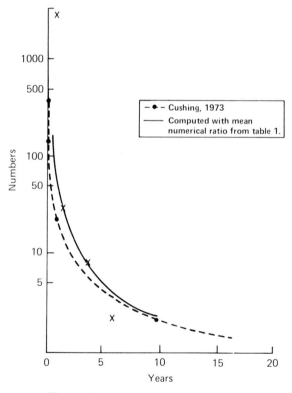

Fig 11 Mean distribution of numbers in year-classes, normalized to age 10 (*ie* 2 fish at age 10).

30

Using Cushing's curve and other available estimates of juvenile age-class strength (*eg* from Andersen and Ursin, 1977), the average numerical ratio of adjacent age-classes with reference to next oldest year-class (N_n/N_{n+1}) (the 'increase factor') was computed and is presented in *Table 1*.

Using the 'increase factors' from *Table 1* and normalizing the number of plaice at age 10 to two individuals, the resulting curve is plotted in *Fig 11* together with Cushing's curve for easy comparison.

The average increase factor should apply only to prespawning and prefishery age-classes. The numbers in fully recruited year-classes can be obtained from actual year-class composition of catches or from long-term mean age-class strengths as described earlier.

Table 1

Numerical ratio of adjacent year-classes (the number presents the factor with which the number of fish in previous year-class must be multiplied to obtain the average number of fish in the year-class under consideration).

Age (year-class)	(0·5)	1	2	3	4	5	6	7	8	9	10
Increase factor	(2·50)	2·10	1·71	1·49	1·36	1·28	1·22	1·17	1·13	1·10	1·075

The factor in *Table 1* is used only as first guess input into an iteration program and is subsequently modified by this program (see description of the computations below). The same factor is used for each species. Species to species differences do appear in final computations due to differences in age composition of the exploited population and turnover rates of biomass.

The fourth basic parameter required for the computation of biomass distribution is the annual turnover rate of the biomass. This turnover rate is usually obtained from ecosystem models, such as DYNUMES and PROBUB (Laevastu and Favorite, 1978 a,b,c). The turnover rate varies from species to species and is normally in the range of 0·5 to 1·2. When turnover rate is not available, computations are made with several plausible turnover rates. Juvenile numbers and biomasses relative to fully exploited age-classes can thus be computed with an iterative method of numerical extrapolation using the increase factor and turnover rate as described below.

The biomass parameters are computed with an auxiliary biomass distribution (BIODIS) model. The model, described below, applies either to age- or year-class, thus from here on we use the term year-class as synonymous to age-class. First we compute the average total mortality per exploited year-class (Z_n) by computing the growth of the previous year-class, adding it to the present age-class biomass (B_n) and subtracting the biomass of the next year-class (B_{n+1}) (see *Fig 12*).

$$Z_n = B_n (1-e^{-g_n}) + B_n - B_{n+1} \qquad (10)$$

where g_n is the growth rate (coefficient) of year-class n.

31

We considered possible nonlinear effects of nonlinear mortalities and growth with age on this calculation. However, these nonlinear effects were found to be small in relation to possible inaccuracies in the basic data used.

The same procedure (*Fig 12*, Formula 10) can also be used for computation of total mortality in juvenile year-classes. However, the number (and biomass) distribution of juvenile year-classes must first be extrapolated, using the assumption that each year-class must produce the next mean year-class and must provide also for predation mortality (which is the main component of 'natural mortality' in juveniles).

The annual growth rate of biomass (G_{an}) (in per cent per year) is computed with the following formula, using the weight at age (W_n and W_{n+1}):

$$G_{an} = [(W_{n+1}/W_n) \, 100] - 100 \tag{11}$$

The corresponding monthly growth rate (G_{mn}) is computed with the well-known compound interest formula: $G_{mn} = (10^a - 1) \, 100$; where $a = \log(W_{n+1}/W_n)/12$.

The corresponding instantaneous coefficients (g) are computed by the well-known conversion: $g = -\ln(1 - G/100)$, where G is the growth rate in per cent.

For the computation of biomass and number distributions, the turnover rate criterion must be satisfied. The latter requires the computation of total mortality of year-classes. The following iterative procedure is used in this computation.

A first guess field of the numbers in juvenile year-classes (N_n) is computed by multiplying successively the next older year-class (N_{n+1}) with the increase factor (f_n) given in *Table 1*, starting with the youngest fully exploited year-class for which empirical data is available:

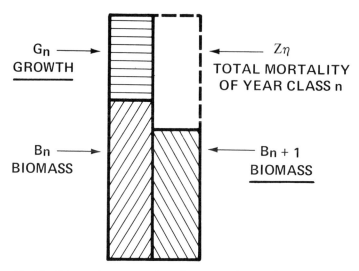

Fig 12   Schematic presentation of mean total mortality of biomass of a year-class (the quantities given with data are underlined).

32

$$N_n = f_n N_{n+1} \qquad (12)$$

Thereafter the first guess of biomass for each year-class is computed by multiplying the number by weight ($B_n = N_n W_n$); then the total biomass mortality in each year-class is computed, assuming it constitutes the difference between the next younger biomass plus its growth minus the next older biomass (see *Fig 12*):

$$Z_n = B_n (1 + 0 \cdot 01\, G_{an}) - B_{n+1}\ \text{or}$$
$$Z_n = B_n (2 - e^{-g}) - B_{n+1} \qquad (13)$$

The turnover rate (T_c) is computed by dividing the sum of biomass mortality (ΣZ) by the sum of the biomass in the different year-classes (ΣB):

$$T_c = \frac{\Sigma Z}{\Sigma B} \qquad (14)$$

The computed turnover rate is compared to prescribed turnover rate. If the difference is larger than a prescribed convergence criterion (*eg* $\pm 3\%$), the following iterative procedure is initiated to converge the computed turnover rate to the prescribed one: Depending on whether the first computed turnover rate was below or above the desired (prescribed) turnover rate, the numerical increase factors (f_n) are multiplied by an adjustment factor (iteration constant) which is slightly above 1 (*eg* $1 \cdot 04$) or below 1 (*eg* $0 \cdot 96$) and the whole procedure is repeated until the computed turnover rate converges to the desired turnover rate (*eg* to $\pm 3\%$). The first year-class is assumed to be comprised of fish from 4 months after hatching to 1 year of age.

The numbers in the exploited year-classes are left unchanged, except the youngest exploited year-class is changed (smoothed) by a small amount to obtain a relatively smooth number/biomass distribution.

The annual mean growth coefficient (g_o) for the whole biomass is:

$$g_o = \overset{n}{\Sigma}\, (G_{an} B_{pn}/100) \qquad (15)$$

where G_{an} is the annual growth rate of year-class n, and B_{pn} is the fraction of the year-class in relation to the total biomass.

The computation of the monthly growth coefficient for the whole biomass or for the juvenile and adult portion of it, is analogous to Formula 15 above, except the corresponding biomass fractions are used in annual (or monthly) time steps. The annual mortality is not simultaneous, but a continuous process.

Therefore, we also need to compute and use various parameters for the deceased portion of the biomass (see *eg* Formula 16 in Section 5.1 Chapter 5). The computation of the growth coefficient for the total population necessitates the use of the deceased fraction of each year-class. This fraction is derived from the total mortality of the year-class. All other desired parameters are computed with simple arithmetical approaches from the input and computed data above.

The reduction of numbers with age is a function of growth rate of the species. The relative distributions of the numbers in year-classes of a slow-growing species (yellowfin sole) and another faster growing species (walleye pollock) are shown in *Fig 13*. By age four numbers have decreased in both species to below 10% of the initial numbers. The faster growing species (pollock) has a somewhat higher percentage of numbers of fish in year-classes 4 to 7 than the slower growing yellowfin sole. This difference is caused mainly by the faster growth of pollock, whereby the juveniles grow faster to the size where the number (and amount) of predators is smaller (*re* size dependent feeding/ predation), thus decreasing the predation in numbers.

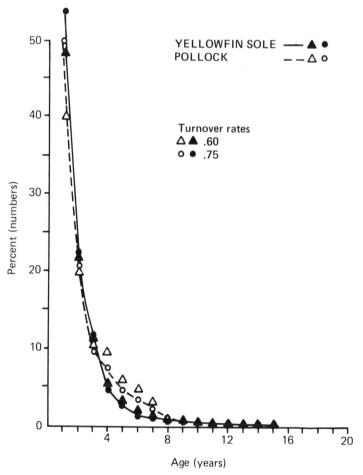

Fig 13 Relative distribution of numbers of fish in different year-classes (in percentage of total numbers from the age of 0·5 year) for yellowfin sole and pollock from the eastern Bering Sea.

34

The distribution of biomass of four species from the Bering Sea is given in *Fig 14* which demonstrates the species to species differences and some peculiarities of biomass distribution. In slow growing species (yellowfin sole, flathead sole, rock sole from the Bering Sea, and Pacific herring) there is one biomass maximum in the first or second year (marked with 'f' on the figures) and another maximum later, when the fish is already sexually mature and being fished (marked with 'h' on the figures). These two maxima are separated by a minimum 'g' which occurs just before or during the sexual maturation of the population. In fast growing species, pollock and rock sole from northern Hecate Strait, there is only one biomass maximum which occurs in the year after the maturation of the population. The population matures also in a shorter period in faster growing

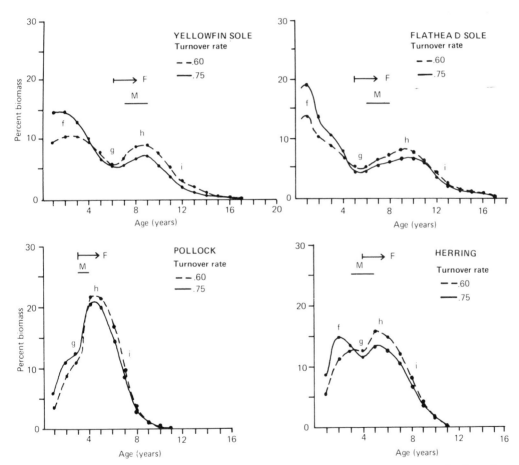

Fig 14 Distribution of biomass with age (at two different turnover rates) of four species from the eastern Bering Sea. (M–period of maturation; F–fully recruited to fishery).

species. The 'minimum' biomass period can also be recognized by a change of biomass increase rate in the faster growing species (marked with 'g' on the figures), although a true minimum does not occur. The decrease of biomass 'i' after the maximum is, in most species, rather rapid and might be related to increase in spawning stress mortality (see Section 5.4 Chapter 5). The effect of different turnover rates on the relative biomass distribution within the population is also shown on *Figs 14* and *15*. *Figure 15* demonstrates the effect of growth rate differences in the same species on the biomass distribution within the population by presenting the same species (rock sole) from two different locations with different growth rates. The vertical bar with the horizontal arrow and 'F' indicates when the year-class has been fully recruited to the fishery.

The biomass distribution computations in prerecruits and juveniles presented in this chapter were based on indirect data. There is an urgent need to conduct further direct, empirical studies on the distribution of year-class strengths of prefishery juveniles of various species, especially in relation to life span of the species and their fecundities. These studies would provide additional verification of the age-dependent mortalities computed with the BIODIS model as outlined above.

This chapter described the methods for computation of biomass distribution and gave some examples. The use of the biomass distribution parameters, computed in this chapter, is described in Section 5.1 Chapter 5 which also explains the dependence of biomass growth rate from the biomass distribution and their changes with time. Section 5.5 Chapter 5 describes the changes of biomass distribution with changing recruitment, and the effects of the variations in fishing intensity on the biomass distribution is described in Chapter 14.

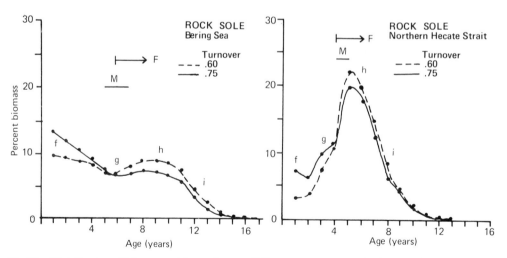

Fig 15 Distribution of biomass with age (at two different turnover rates) of rock sole from the eastern Bering Sea and northern Hecate Strait. (M—period of maturation; F—fully recruited to fishery.)

5

Major processes in the fish component of a marine ecosystem and their simulation

The quantitative numerical simulation of a fisheries ecosystem consists essentially of computation of changes of abundance and distribution of species biomasses with time. A given ocean region is taken as space unit, which can be further divided into many sub areas by designing a computational grid, where computations are done at each grid intersection (grid point) and in each time step (usually each month). The following processes are computed on each species biomass:

——Growth as affected by various environmental factors and by availability of food.

——Food uptake, computed by food items, which will also give predation mortality.

——Other mortality components, such as spawning stress mortality and senescent and disease mortalities.

——Spawning (and larval recruitment).

——Migrations from grid point to grid point.

——Fishery (fishing mortality).

The 'model' and computation formulas are summarized in Chapter 6. Some readers might prefer to review Chapter 6 before searching for details of the quantitative simulation of various processes in this chapter, whereas some readers would prefer to obtain the background of the essential processes from this chapter before reviewing Chapter 6.

In this chapter some essential processes in the ecosystem are quantitatively reviewed and summarized to the extent necessary to explain their simulation in ecosystem models. In the first section we summarize the factors affecting the growth of biomass and present the formulas for computation of the changes of growth. Food requirements of the biomass for growth and maintenance are summarized in the second section. The third section describes the simulation of the spatial and temporal changes of food composition. Section four presents computation of and new results on spawning stress mortality. Simulation of recruitment in biomass-based model is described in section five. The final section, six, describes the simulation of migrations in ecosystem models and summarizes the effects of migrations in this system.

5.1 Growth of biomass and factors affecting it

Growth of biomass is one of the basic processes in the ecosystem, thus the growth rates of fish populations are among the more important parameters in ecosystem simulation. Paloheimo and Dickie (1965) have summarized the processes of growth of individual fish. The growth rate (in terms of weight per unit time) is highest in larvae and juveniles and decreases considerably with age. Growth varies with seasonal changes of temperature and with the availability of food. The changes in growth rates of most species with latitude (and general location) as well as with time in a given region are usually well known (examples of annual growth rates of rock sole in two different locations are shown in *Fig 16*). Such differences in growth rate will affect the distribution of cohort biomass within a population (Chapter 4).

For a fast growing species the growth rate in the first and second year can be several hundred percent of the mean biomass of the age-class. By the fourth year the growth rate decreases in most species to about 40% per year and about a year after maturation the annual growth rate decreases to 20% and below. The main inflection in growth curves related to maturation has been pointed out earlier by Paloheimo and Dickie (1965). About three years after maturation there is another characteristic decline in growth rate (marked with 'a' in *Fig 16*), the significance of which is not fully clear. The lower growth rates after maturation are probably caused by physiological changes, such as energy requirements for sex product development and difficulties in obtaining adequate amounts of food.

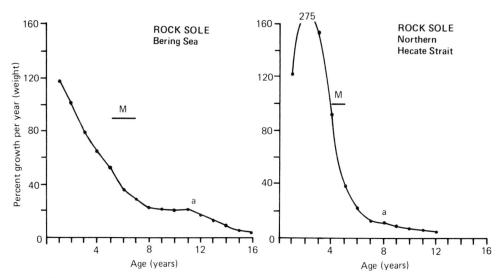

Fig 16   Annual growth rates (in percentage of weight per year) of rock sole from the eastern Bering Sea and northern Hecate Strait. (M–period of maturation.)

38

The growth rate of individual age-classes is not used directly in biomass-based ecosystem models but is used for computation of the growth rate of the population biomass. This growth rate is also a function of biomass distribution with age. The biomass-based ecosystem models use a monthly growth coefficient (g_0) which is computed by multiplying the annual growth rates of the species (G_{an}) (derived from empirical data on weight-at-age) with the fraction of the biomass in a given year-class ($B_{pn}/100$) (Formula 15).

The growth rate used in simulation computations is the mean of the 'whole population' growth rate (g_{tot}) and of the 'deceased population' growth rate (g_{dec}) (see *Table 2*).

$$g_0 = 0.5\,(g_{tot} + g_{dec}) \tag{16}$$

As an example of biomass growth rates, a summary of biomass growth coefficients for rock sole from the eastern Bering Sea and from northern Hecate Strate is given in *Table 2* as monthly compound percentage of growth (analogous to compound interest rate). The biomass growth coefficients are also dependent on the turnover rate of the whole biomass of the population. (Two turnover rates which are representative for flat-

Table 2
Distribution of growth, biomass, and numbers between juvenile and exploitable population of rock sole from the Bering Sea and from northern Hecate Strait.

	Subject	Bering Sea turnover rate		Hecate Strait turnover rate	
		·60	·75	·60	·75
Monthly growth coefficient (%)	Whole population	3·25	3·64	3·20	3·75
	Juvenile population	5·22	5·34	8·69	8·64
	Exploitable population	1·73	1·74	2·21	2·27
	[1]'Deceased' population	4·19	4·62	5·07	5·44
Biomass (%)	Juvenile population	43·6	52·8	15·3	23·3
	Exploitable population	56·4	47·2	84·7	76·7
Numbers (%)	Juvenile population	92·2	94·8	77·1	86·7
	Exploitable population	7·8	5·2	22·9	13·3
Total Mortality					
Biomass (%)	Juvenile population	66·8	75·3	32·9	49·8
	Exploitable population	33·2	24·7	67·1	50·2
Numbers (%)	Juvenile population	97·8	98·5	92·2	96·1
	Exploitable population	2·2	1·5	7·8	3·9

[1]Computed growth rate of the part of the population which has died during the year (mostly from being eaten).

fishes were used for the data in *Table 2*: 0·60 and 0·75) The most pronounced difference in the growth rates of rock sole in the two locations is in the juvenile population which grows much faster in more southern latitudes (northern Hecate Strait) than in the Bering Sea. The theoretically computed growth rate of that part of the population which has died (been eaten) during the year, the 'deceased' population, is normally higher than the mean for the 'whole population' because the greatest removal of the population biomass occurs in juveniles through predation mortality where the growth rates are higher. If several species are combined into an ecological group, the growth rate of this group must be a function of the relative abundance of the species in the group.

The growth rate varies seasonally in most species in medium and high latitudes. This seasonal variation can be reproduced either as a harmonic function of time or as a function of water temperature. A simple harmonic function is:

$$g_t = g_0 + g_m \cos(\alpha t - \kappa_g) \tag{17}$$

where: g_t is the monthly growth rate for month t, g_0 is the monthly mean growth rate, g_m is half of the magnitude of annual change of growth rate, α is the phase speed (depends on time step; 30° for monthly computations), t is time (months), and κ_g is the phase lag.

One of the main causes of the seasonal changes of growth is the change of temperature of the water. Thus the growth can be made a function of temperature (either surface or bottom temperature, depending on whether the species is pelagic or demersal). The growth rate maximum can be at different temperatures for different species and in some species growth ceases at 4°C. Nearly every species in higher latitudes has a specific optimum (acclimatization) temperature and in some species the optimum temperature can have a wide range.

In a comparative study of the temperature effect formulas, the following has been found best (Krueger, 1964):

$$g_t = g_0 e^{\left(\frac{1}{T_0} - \frac{1}{T}\right)} \tag{18}$$

where T_0 is optimum (acclimatization) temperature for the species and T is actual temperature. This formula (18) permits the assignment of optimum temperature for each species/group of species and does not give abnormally high growth at high temperatures as some simpler formulas do. An example of the effect of temperature on the growth rate is given in *Fig 17*.

A linear relation between growth in weight and food intake has been found to be applicable to the growth of fish (Jones and Hislop, 1978). If a fish obtains full daily ration of food, no changes in the growth coefficient are necessary. If, on the other hand, growth is made a harmonic function of time, the food requirement must also be made a harmonic function of time.

To account for the effects of changes in growth caused by temperature on food intake, food requirements must be computed in two steps: food requirements for maintenance and food requirements for growth.

As the availability of food affects the growth of the fish population, it is imperative to test the food availability and possible food substitution at each time step (and grid point) in the simulation model. If full substitution of food which is absent or short in supply at a given location and time is not possible, partial starvation is considered to occur. There are several subjective ways to compute the effect of starvation on the growth. The actual (complex) method used in simulation is best described in the model documentation where the FORTRAN code is given. The following description serves only as an example for computation of food composition change and starvation. First, the fraction of biomass of a given species consumed in the ecosystem in the previous time step (month) (R_{t-1}) is compared to the maximum allowable fraction $(R_{o,j})$ (predetermined), *eg* 4% of total species biomass per month. If the actually consumed fraction in the previous time step exceeds the allowable fraction, then the prescribed mean fraction of this species in the food of the species under consideration $(p_{i,j})$ is decreased

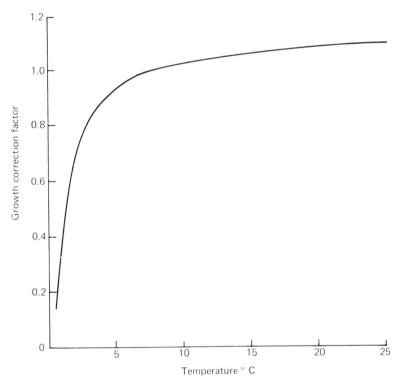

Fig 17 Effect of temperature on growth rate (Formula 18); acclimatization temperature 8°C.

41

$p_{i,j} = (R_{o,j}/R_{t-1})\, p_{i,j}$. The new food composition for species i is summed and the missing fraction of the food requirement is divided between those food items which had an ample supply in proportion to their occurrence in the mean food composition (prescribed at the start of the computations) and serves as a vulnerability index. However, if the missing fraction is large (*eg* in excess of 20% of food requirements), part of this missing fraction is recorded as starvation, *ie* subtracted from normal food requirement.

The effect of starvation on growth rate is assumed to be linear:

$$g_t = g_o - \left(\frac{S_i}{R_i} g_o \right) \tag{19}$$

where S_i is the shortage of food at a given time step and location and R_i is the full food requirement.

Biomass distribution with age changes with the intensity of the fishery. Thus, the growth coefficient increases with the increase of fishing (rejuvenation of population, Dementjeva, 1964). Detailed computations of the change of the growth coefficient due to a change in fishing intensity are not practicable, as the exact age composition of the exploitable biomass is seldom accurately known. Instead an approximate 'correction' with the following empirical formula could be made.

$$g_t = g_o\, a\, \frac{F_n}{F_o} \tag{20}$$

where a is an empirical constant, F_o is the old fishing mortality coefficient and F_n is the new fishing mortality coefficient.

Variations in recruitment affect the growth rate of the population. The recruitment (and spawning) can be considered either as a continuous function of biomass with time or a semicontinuous function of biomass with time. The recruitment in biomass-based models is described more fully in Section 5.5.

If one considers a continuous recruitment to all age/size groups and assumes that there are no exceptionally strong or weak year-classes of postlarval juveniles, recruitment would be proportional to the biomass present. Variations in postlarval recruitment would be depicted in biomass-based models in the variations of the growth coefficient of the species biomass (if the species is treated as one unit). This is shown in *Fig 18* where high growth rate in postlarval juveniles (dotted line) and an increase in biomass of these postlarval juveniles, would result in increased overall mean growth rate coefficient for the species. A strong year-class of older fish would lower the mean growth rate coefficient.

On the other hand, large spawning biomasses are known to produce proportionately smaller year-classes and small spawning biomasses are known to produce proportionally large recruitment (year-classes). Therefore, the effect of recruitment on growth rate can be regulated in biomass-based models, making the biomass growth rate coefficient inversely proportional to biomass present.

$$g_c = g_t \sqrt{B_{e,i}/B_{i,t-1}} \tag{21}$$

where $B_{e,i}$ is the equilibrium or mean biomass of species i. This computation can be done in the models in a prognostic mode after the determination of the equilibrium biomasses (see further Section 5.5).

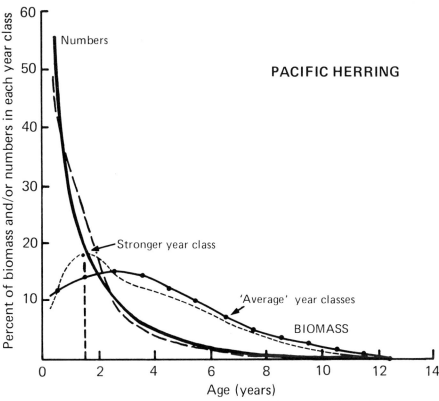

Fig 18 Distribution of biomass of Pacific herring within different year-classes (percentage of total), with average year-classes and with a stronger than normal year-class.

5.2 Food requirements of fish biomass

Trophodynamics (food uptake and utilization) has not been used in single-species population dynamics. Many diverse treatments of feeding, food requirements, uptake and utilization by fish are found in the literature. In this chapter we select and describe those fish trophodynamic aspects which have been used in biomass-based ecosystem simulation. We present empirical formulas for computation of food uptake and its partition for different functions in the biomass dynamics and present examples of numerical values of coefficients and describe their dependence on various biotic and abiotic factors.

43

We define food requirement as mean saturation uptake of food biomass (*ie* food eaten in optimum conditions with unlimited food availability) per unit biomass of a given stock per unit time. The food requirement has often been termed ration. The food requirement can be partitioned between food requirement for growth, for maintenance of activity (maintenance ration at which biomass neither decreases nor increases), and for sex products (gonad) development. The amount of food actually eaten is often less than the food requirement and is computed in the ecosystem for each species, time step (month), and location.

The information in the literature on the food requirements of fish is diverse and at times seemingly contradictory, thus requiring serious evaluation as to reality and applicability in ecosystem simulation models. Most empirical information pertains to fish culture and fast-growing fish. Its applicability to natural populations is often indirect.

Some reports pertaining to trophodynamics have used the energy content of food as well as that of predators as a basis for calculations, rather than biomass. Although the caloric value serves some useful purpose in fish culture, the reasons for the use of caloric values in the studies of wild (natural) ecosystems is unclear and artificial. Caloric value of prey items is known only superficially. Stomach content has not been measured by its caloric content but by volume. Furthermore, the 'caloric value' of fish varies with age, season, and location. (Some reports give the 'natural food' caloric contents as 330 kcal/kg, others report a value of 500 kcal/kg).

Most research on food requirements by fish has expressed these requirements as functions of maintenance and growth requirements, plus food requirements for development of sex products.

$$R = aG + bW + cW \qquad (22)$$

where: R is the total food requirement per unit time; aG is food requirement for growth [G being growth of biomass and a is a coefficient (food coefficient for growth)]; bW is maintenance ratio, which is a function of biomass (weight) (W) and basic maintenance requirement per unit time (b, *eg* 0.7% of body weight per day). The factor b is also a function of the activity of fish. The last term (cW) is the food requirement for sex product development which can be (and often has been) included in the first term (aG) under certain conditions.

Some examples of the various estimates of food requirement of fish are given below. The food demand for growth versus demand for sustenance, is estimated in fish culture as $\approx 1:2.3$. An often used estimate of food conversion is that about 30% of food fed is converted to tissue (*ie* 'food coefficient' about $1:3$). Typical results with high-caloric food in fish culture is that about 3kg food is required to produce 1kg fish (Halver, 1972). Last (1979) reported a conversion efficiency of 10% to 20% in fish and that juvenile fish can take up food to 28% of body weight daily (BWD). Most commonly used annual food consumption is 3 to 4 times the mean biomass (standing crop) per year which after conversion is about 1% body weight daily.

44

For about half a century the marine ecosystem has been assumed to consist of definite 'trophic levels' with a trophic level 'conversion factor' of 10:1. In reality, trophic levels are undefinable in a real ecosystem as briefly described in Chapter 4 and in the event some definition would be possible, most fish change the 'trophic levels' throughout their life history, throughout the year, and from area to area due to space and time variation of food composition.

Daan (1973) found from stomach content and food uptake studies that cod in the North Sea required an average 0·5% to 1·5% body weight of food daily. Daan's data converted to a mean value for cod biomass is about 0·75% of BWD, which includes food for maintenance as well as for growth. Tyler and Dunn (1976) estimated that flounders require 1% BWD for maintenance, and other authors have estimated the total food uptake by fast-growing salmon to be as high as 2% BWD. After considerable review of the literature and after experimentation with a single food requirement (defined as fraction of body weight daily), we decided to use the following formula for food requirement in the ecosystem simulation:

$$R = bW + aG \qquad (23)$$

where R is the total food requirement per unit time, b is food requirement coefficient for maintenance in terms of percent BWD, W is biomass (weight), a is food requirement coefficient for growth (including sex products development) in terms of growth, and G is growth of biomass. The coefficients a and b vary from species to species. Although the mean value for the coefficient b is about 0·75, if we use a single food requirement factor, in the above formula b is about 0·55 (varying from species to species). Furthermore, it is also a function of temperature as is the growth (see Formula 24). The coefficient a is about 1·2, varying from species to species. A numerical example would be as follows:

Biomass, (monthly mean) 100 (kg/km²)

Growth – 6% per month	= 6 (kg/km²)
Food requirement for maintenance at equilibrium temperature and with 0·55% BWD ($\times$ 30 days)	= 16·5%
or in terms of weight	= 16·5g
Food requirement for growth (6 × 1·2)	= 7·2g
Total food requirement	= 23·7g
Food requirement in terms of BWD (growth + maintenance)	= 0·79% BWD

45

Amount of food eaten is a function of the availability of suitable food items (by size and by species) at a given time and location. Thus a check has to be made in the simulation at each time step and location (grid point/region) whether sufficient food is available to a given species (or age group, if so divided) so that the fish can obtain the optimum ration. If a given food item is not present in sufficient quantity (or has been 'overconsumed' in the previous time step), possibilities of substitution with other food items, specified in the diet for the given species, must be attempted. If, however, a full substitution is not possible, it must be assumed that partial starvation occurs, the measure of which is the difference between full ration required and that which can be taken. The growth rate of the biomass must consequently be adjusted if partial starvation occurs (see further Section 5.1).

There are some other consequences of partial starvation. Tyler and Dunn (1976) found that partial starvation of flounders in Passmaquaddy Bay occurred and that the fish seemed to sacrifice egg production to maintain body weight. Flüchter and Trommsdorf (1974) also found that malnutrition was the reason for incomplete development of eggs in the common sole.

Many environmental factors influence the behaviour of fish and its feeding. One of the factors is the seasonal temperature. Both too high and too low temperatures cause the lowering of the feeding rates. Different species have different optimum temperatures at which feeding is at maximum. It is also known that 'abnormal' temperatures hinder fish from migrating into areas where the food might be abundant.

In high latitudes growth varies with season. This seasonal growth rate change can be simulated either with a harmonic formula or with seasonally changing temperature (see Section 5.1). As growth rate is related to food requirement, the latter must also be made a harmonic function or function of temperature. In our simulation, growth is a function of temperature (which varies with seasons) and, because food requirement is a function of growth, it becomes also a function of temperature, thus varies seasonally.

The maintenance ratio is assumed to be a function of the activity of fish, and both the activity of fish and metabolism are functions of temperature. As temperature is taken into consideration for growth computations, it can be neglected for computations of food requirements for growth if Formula 23 is used. Quantitative empirical data for the effect of temperature on maintenance ration are scarce. Jones and Hislop (1978) presented the following formula: maintenance requirement $= 0.0080W\,e^{0.081T}$ where T is in °C and W is in g. In our simulation we have assumed that the effect of temperature on metabolism (and maintenance food requirement, which reflects also the activity of fish) is the same as for growth; we have thus used the following formula:

$$R = bWe^{\left(\frac{1}{T_0} - \frac{1}{T}\right)} + aG \tag{24}$$

where R is total food requirement per unit time, b is food requirement for maintenance in terms of decimal fraction of body weight per unit time (usually given as body weight daily), varying from species to species, W is biomass of the species, T_0 is 'acclimatization temperature' and T is actual temperature, a is an empirical constant (about 1·2, varying somewhat from species to species), and G is growth of biomass per unit time.

As growth varies with age (size), the total food requirement will vary with age (size) of the fish. As the growth rate of a population changes with its age composition, the 'rejuvenation' of a population would result in higher food consumption per unit biomass. However, 'rejuvenation' usually involves a decrease of total biomass of a stock. Thus, it seems that if a stock size does not vary within very wide limits, its food consumption remains quasi-constant despite rejuvenation.

The growth rates and food requirements are main determinants of the 'equilibrium biomass' (the unique solution in the PROBUB model, see Chapter 10). Thus, to obtain a 'minimum equilibrium biomass', highest plausible growth rates and lowest plausible food requirements should be used in the model for this purpose.

The next section will describe the simulation of the amounts of food actually eaten and the occurrence of partial starvation. Furthermore, we will also describe the simulation of space and time variation of food composition.

5.3 Composition of food, its spatial and temporal changes, and predation mortality

The composition of food of any species in any specific time and location depends on the availability of preferred food species of proper size. The average composition of food items of individual species is specified with input data in the fish ecosystem simulations. This average composition of food is obtained from the literature or from other data sources. The prescribed food composition serves as an index of vulnerability to predation; the actual composition of food is computed in each time step and location considering availability of individual food items and using the initial average food composition as an index of vulnerability to predation.

Size-dependent feeding dominates in the marine ecosystem. The predator weight to prey weight ratio varies somewhat from species to species. The mode of this ratio is, in average, somewhat larger than 100. The corresponding length ratio is about 10. Due to age dependent growth and mortalities, the availability of food of proper size decreases with increasing predator size so that larger predators must, in some conditions, feed on smaller than optimum size prey.

The predation is highest on larvae and juveniles; it decreases with increasing size of the fish. When the composition of biomass by age/size and age-dependent predation mortality are computed for each species, one can arrive at a realistic limit of the amount of predation which is possible from a given species per unit time. This limit varies usually from 3% to 10% of mean biomass of prey per month but can be higher during a few months after spawning, because of the presence of vulnerable eggs and larvae. This 'maximum allowable predation' must be prescribed for each species in the model (see further Section 5.1). The maximum fraction of the biomass of a species available to predation is directly proportional to the growth rate; the faster the growth of larvae and juveniles, the faster they pass the predation vulnerable phase. Ricker and Foerster (1948) and Beverton and Holt (1957) also pointed out the effect of rate of growth on the time of passing the most vulnerable predation age (size). A general relation also exists between fecundity and predation mortality with more vulnerable species having higher fecundity.

Extensive quantitative food composition studies of marine fish are few. Additional information on average food composition must thus be obtained by considering the feeding regime (pelagic or demersal), the diurnal mobility of the species, and their seasonal migrations. As is known, fish do not feed continuously; several species migrate diurnally between the surface layers at night where the pelagic food may be abundant and the near bottom layers where they usually spend the day.

Relatively few data are available on the feeding behaviour of fish, especially on the strategy of searching for food. Feeding behaviour varies with the changing availability of the quantities of food, as well as with the availability of preferred food. Generally fish aggregate when and where the food is abundant and disperse when the food is scarce; no quantitative criterions have, however, been established for this behaviour.

The prevailing feeding in the marine fish ecosystem can be characterized as opportunistic, availability-dependent, 'regime defined' (ie pelagic or demersal), and size-dependent, with some preference (selection) for food items. Availability (density) dependent feeding implies that if one preferred food item becomes more abundant, its consumption would increase. On the other hand, if there is a lack of a preferred food item, substitution with a similar sized prey, either lower or higher on the average food composition list, might be made. If no substitution is possible, partial starvation will occur which would affect the growth rate.

Another important aspect of food composition that has been relatively little recognized is cannibalism. There are few examples of it in the literature. Daan (1976) found that 'The most important food item by weight is, for the herring, its own off-spring, but in numbers it is surely the fish eggs'. Mito (1972) gave the maximum cannibalism in pollock in the Bering Sea by the size of pollock: 24-37cm – 71%; 42-55cm – 84%; > 55cm – 100%.

There are various ways to simulate variable food composition of a given species in the marine ecosystem. The following is one of the examples used in the PROBUB model.

R_0 is the 'maximum fraction' of any food item allowed to be consumed at a given time step (eg 4% of available species biomass per month). R_{t-1} is the fraction actually consumed in the previous time step.

If $R_{t-1}/R_0 < 1$

then $p_{(i,j,t)} = p_{(i,j,o)} = p_{vj}$ \hfill (25)

where: $p_{(i,j,t)}$ is the decimal fraction of food item j in the food of species i at the time step t, $p_{(i,j,o)}$ is the above quantity as prescribed with input (*ie* the average composition of food), and p_{vj} = the unchanged composition ($= p_{(i,j,o)}$).

If $R_{t-1}/R_o > 1$
then the changed decimal fraction (p_{mj}) is:

$$p_{(i,j,t)} = p_{(i,j,o)} R_o/R_{t-1} = p_{mj} \tag{26}$$

and the missing decimal fraction of food (p_m) is:

$$p_m = p_{(i,j,o)} - p_{(i,j,t)} \tag{27}$$

After the determination of equilibrium biomasses with the PROBUB model, the food composition as determined at the equilibrium condition will be changed as the food density changes, *ie* the term R_o/R_{t-1} will be replaced with B_t/B_e – where the B_t is the actual biomass and B_e is the equilibrium biomass (both referring to the food item).

The sum of unchanged food composition (P_s) is:

$$P_s = \overset{j}{\Sigma} p_{vj} \tag{28}$$

and the sum of the missing food (P_m) is:

$$P_m = \overset{j}{\Sigma} p_{mj} \tag{29}$$

If P_s is > 0.85 (this is a subjective decision and an example only), then the adjusted decimal fraction ($p_{(i,j,t,a)}$) is:

$$p_{(i,j,t,a)} = p_{(i,j,t)} + [(p_{(i,j,t)}/p_s) P_m] \tag{30}$$

If P_s is smaller than 0.85, then some starvation can be assumed to occur and some subjective decisions must be made about what size of fraction of the food (V_p) could be taken from food items present and how much 'starvation' (S) should occur (V_p and S are in decimal fraction).

If P_s is < 0.85 but > 0.50; then $V_p = 0.80$; $S = 0.20$
If P_s is < 0.50; then $V_p = 0.68$; $S = 0.32$
Final food composition (*ie* fraction taken) ($p_{(i,j,t,f)}$) is

$$P_{(i,j,t,f)} = p_{(i,j,t)} + (V_p p_{(i,j,t)}/P_s)P_m \tag{31}$$

and

$$S_n = S P_m \tag{32}$$

It can be argued that predation might control recruitment (Rothschild and Forney, 1979). Consequently, if predation in a given area and year is below average, an increased survival and increased biomasses would result.

The apex predators (marine mammals and birds) have considerable mobility and can thus search and find concentrations of preferred food more easily than many fish species. Thus they are more selective with respect to food than are other components of the ecosystem. Their food composition varies in the model only from one region to another.

Quantitative food studies of marine fish are scarce, as mentioned earlier, which might be partly due to the fact that the results of these studies were not considered fully in single-species dynamics. However, as trophodynamics is an important part of the ecosystem, it becomes imperative that fish food research be assigned a high priority in future research, with special emphasis on large predatory fish and also in respect to the space-time variation of predation pressure on larvae and juveniles. These studies might also add knowledge on the predation avoidance in juveniles. Many general predation avoidance and minimizing behaviours and factors are qualitatively known. These factors vary considerably from species to species and from area to area, such as the migration (and transport) of juveniles into the environments with fewer predators, long spawning times of stocks (resulting in lower larval densities), spawning several times over several months (*eg* California anchovy), *etc*. Unfortunately none of these factors can be well quantified to serve as indices for survival (and recruitment) variations. Some of these specific behaviours of juveniles and their effects can, however, be studied at least semi-quantitatively with ecosystem simulations.

5.4 Senescent, disease, and spawning stress mortalities

Mortalities from natural causes have, in general, been 'guessed at' quantities in fisheries population work in the past. The mortalities in the first year (mainly predation mortalities) can be about 50% per month and more. Cushing (1976) found that the mortality of plaice larvae in the first months is about 80% per month. Furthermore, Cushing estimated that during the first year the increment of weight of a population is on the order of 10^5 but the decrement of numbers is 10^4, so that the gain of the biomass can be one order of magnitude (fish larvae can grow 10% a day).

Lett and Kohler (1976) have been among the first to recognize the importance of predation mortality which, tempered by available food density, is assumed to be a major population stabilizing and fine-tuning mechanism for year-class strength formation.

Mortalities are classified by various causes in the ecosystem simulation:

——Predation mortality – in most cases this is the largest mortality component (the main component of the earlier used term 'natural mortality'). This component, which includes mortality from cannibalism, is computed in detail in the ecosystem simulation with the trophodynamic approach (see Sections 5.3 and 5.6).

——Fishing mortality – determined from catches and from exploitable biomasses present; its computation and use in biomass-based models is described in Chapters 14 and 15.

——Spawning stress and senescent mortalities are usually considered together. There is an upward trend of mortality in older animals which has been termed senescent mortality; lately it has been considered that in fish the main component of this senescent mortality is spawning stress mortality. (This mortality is discussed in greater detail below.)

——Residual 'natural mortality' – which may be caused by diseases and infestation by parasites (disease mortality) and whose magnitude is at present unknown but can be assumed to be small under 'normal' conditions; a small natural mortality coefficient is therefore prescribed for all species to account for this mortality. This coefficient is assumed to be higher in species which are not subject to fishing to account for increased senescent mortality.

——'Special mortalities' – the mortality from exceptionally cold conditions is treated in the ecosystem simulation by increasing the natural mortality subjectively in areas and times when especially cold anomalies occur. Some 'starvation mortality' is assumed to occur if computed partial starvation is above a given predescribed level. The natural mortality coefficient is also slightly increased (at times and locations) when this occurs.

The numerical mortality rates for individual age-classes of the exploited part of the population can be computed directly using the long-term mean numerical age composition of the fully exploited part of the population (*Fig 19*). These total mortalities are plotted on *Fig 20* as a percentage of mortality (in numbers) from the previous, fully exploited year-class. The beginning of the year in which the species is fully recruited to the fishery is marked with a vertical bar (F) and the maturation period with a horizontal bar (M). Before and during the maturation period, the numerical mortality (in numbers) remains in general below 15% and lower in species other than herring. As herring, and partly pollock, are important forage fish for marine mammals in the Bering Sea, their mortalities during maturation are somewhat higher than in other fish (15 to 20%). The 'prematurity mortality' in the exploitable part of the population consists mainly of long-term mean fishing mortality plus a very small mortality from diseases and grazing (predation) by mammals (the latter only in pelagic and semipelagic species – herring and pollock). In most species the total mortality during this period is essentially fishing mortality, as the small additional mortality from diseases and predation can be ignored (Dementjeva, 1964). If no additional mortality component arises in a subsequent year, this long-term mean fishing mortality should remain approximately constant in each year-class. (Note that the numerical mortality in *Fig 20* is computed relative to previous year-class numbers and would thus remain constant if the mortality rate did not change with age due to changes in vulnerability to fishery.) A horizontal line of graphically estimated fishing mortality (M_f) has been drawn on *Fig 20*. A possible error of about 12% arises in estimating the relative age-class strength of the first fully recruited year-class which would affect only the first value of mortality on *Fig 20*. The prematurity (mainly fishery) mortality in some Bering Sea fishes has been roughly estimated at 6% per year in yellowfin and flathead sole, 10% in rock sole, 15% in pollock, and 19% in herring; some predation mortality by mammals on the latter two species is included.

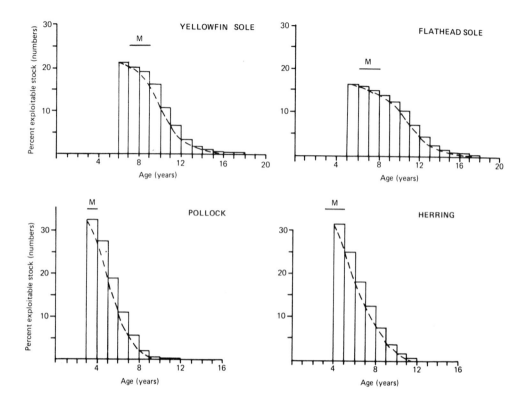

Fig 19 Long-term mean age frequency distribution of fully recruited year-classes (in percentage of numbers) of four species from the eastern Bering Sea. (M–maturation period.)

After the cohort has reached about 80% maturity, the mortality shows an increase of about 10% per year for the next 4 to 5 years and thereafter undergoes irregular fluctuations. These irregular fluctuations (and corresponding mortality values) are somewhat uncertain, due to the smallness of the sample (too few individuals alive) and might be caused also by a positive bias in the formation of mean age-class strength induced by stronger year-classes in area of the summation of older year-classes where variation is bounded by zero. A sloping dashed line has also been drawn on *Fig 20* (designated $M_f + M_s$), starting from the estimated fishing mortality level in the year when the cohort reaches about 80% maturity. This line has a slope of 10% increase of mortality per year on each figure and coincides reasonably with the average slope of empirical mortality data during the first 4 to 5 years. However, there are some small displacements either toward younger or older years. The reasons for these displacements (and possibilities to achieve better agreements between the 10% trend line in mortality increase and corresponding empirical data) are most likely in the uncertainties in the preliminary data on the maturation of the populations. A 1-year earlier, full maturation

52

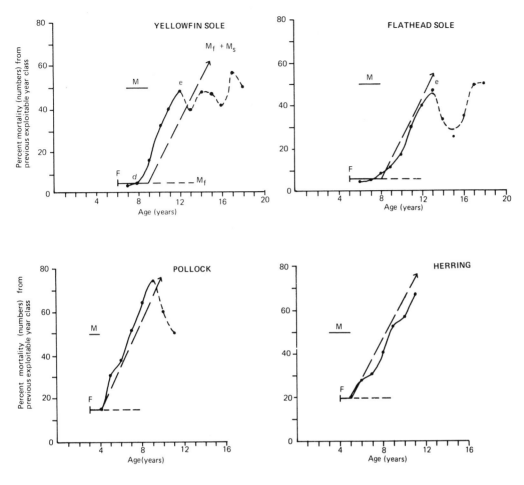

Fig 20 Percentage of mortality (in numbers) from previous fully recruited year-class, for four species from the eastern Bering Sea (explanations see text).

of yellowfin sole and 1-year later maturation of flathead and rock sole from the Bering Sea would produce closer fits between the trend lines and empirical data. Furthermore, year-to-year variations in growth rates and correspondingly earlier or later maturation can occur in most marine species (Dementjeva, 1964).

The increase of apparent mortality with age might be partly due to decreased vulnerability of larger fish to the gear and partly due to different distributions (*eg* larger fish in deeper water). Although the above-mentioned factors might contribute to the apparent increased mortality with age, it is unlikely that they are the only factors (especially considering that the mortality increase starts at and after the maturation and is nearly the same in all species studied), despite the fact that their biological characteristics and modes of life are quite different.

Beverton (1963) and Cushing (1976) have also noted this increase of mortality with age and have attributed this to increased senescent mortality. Cushing (1976) found the senescent mortality increase in North Sea plaice to be 8·9% per year, using a quite different approach.

As the mortality starts to increase after maturation, we believe that the main cause of this mortality increase is spawning stress. The spawning stress mortality is well known and pronounced in anadromous and catadromous fish. In marine fish it was quantitatively considered first by Andersen and Ursin (1977). The present data indicate that the total mortality of any species in an exploited year-class where predation mortality can be ignored, can be computed with a relatively universal formula, which is in essence Gompertz's Law (Beverton, 1963):

$$Z_n = F + n_s M_s \qquad (33)$$

where Z_n is total mortality of the year class (in percentage of numbers from previous exploitable year-class); F is the basic fishing mortality (5 to 20% in our examples), n_s is the number of years after the cohort has reached 80% maturity, and M_s is a spawning stress mortality which increases 10% per year. If the spawning stress mortality follows a linear increase with time, and if the fishing mortality rate can be estimated from other data, the above approach can be used to compute the mean total mortality for the exploitable part of a population if the age composition of this population has been ascertained.

$$Z = \overset{n}{\underset{}{\Sigma}}\, a_n \,(F + n_s M_s) \qquad (34)$$

where Z is the total mean mortality of the exploitable part of the population expressed either as a percentage or converted to an instantaneous mortality coefficient and pertains to numbers or to biomass, depending on whether the decimal fraction a_n of a year-class is given on the basis of numbers or biomass. The remainder of Formula 34 in parenthesis is the same as Formula 33, where M_s is expressed as the 'universal average spawning mortality increase rate' (10% per year-class or 0·096 as an instantaneous rate per year).

One of the consequences of spawning stress mortality was mentioned by Dementjeva (1964), although she did not recognize spawning stress mortality *per se:* 'The earlier a fish reaches maturity, the shorter is its span of life.' Dementjeva furthermore suggested that the increase of growth and earlier maturation leads to rejuvenation of stocks, *ie,* to an increase of the biomass in younger year-classes and corresponding decrease in the abundance of the older specimens of the fished stock. The intensity of fishing also plays a role in this process.

Figure 21 presents the computed total mortality of the population biomasses of four species with age, expressed as percentage of mortality of the corresponding annual mean biomass of each year-class. There are a number of characteristic features in this total mortality distribution with age which are found in all species presented on these figures. First, there is high mortality of biomass of young year-

54

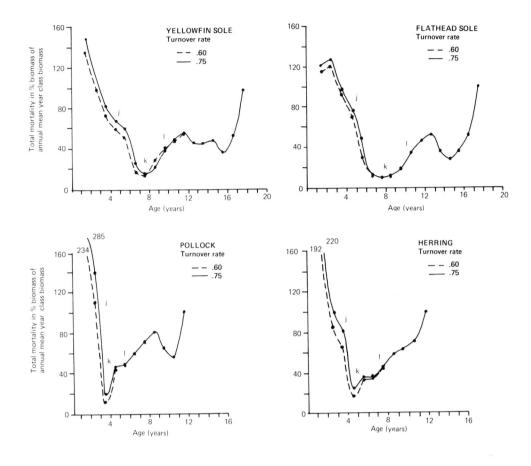

Fig 21 Distribution of total mortality with age (in percentage of biomass of the annual mean biomass of given year class) for four species from the eastern Bering Sea.

classes of all species. This juvenile mortality declines rapidly with age – the more rapid the decline, the faster the growth of the species. The high mortality in larvae and juveniles is mainly due to predation. When the fish reaches the size category where the number of predators who are able to catch and/or swallow it is small, the mortality reaches the lowest value (marked with 'k' on the figures). During this low predation mortality period, the main mortality is fishing mortality in commercially caught species.

As soon as the cohort reaches full maturity, the spawning stress mortality sets in as discussed above (marked with 'l' on the figures). Obviously, the numerical values of the spawning stress mortality increases are different in terms of biomass (*Fig 21*) than in terms of numbers (*Fig 20*).

55

5.5 Recruitment in the biomass-based ecosystem simulation

The recruitment problem in fisheries prediction and management has been of paramount importance. Unfortunately relatively little progress has been made over the decades to explain quantitatively the processes affecting it and to make recruitment quantitatively predictable. The term recruitment itself is ambiguous; it applies to the recruitment of larvae as well as recruitment to the exploitable portion of stock. There is a variable number of years between these two 'entry points' and several processes reducing the numbers and increasing the biomass occur.

The early single-species population dynamics approaches applied an assumed stock recruitment relation. The Ricker stock-recruitment curve is dome-shaped, implying stock-dependent mortality, which can only be complete cannibalism (Harris, 1975) although this scarcely occurs with such intensity. Thus Ricker's stock-recruitment curve cannot be applied to multi-age stocks and the Beverton-Holt curve should not be applied to gadoid stocks (Cushing, 1976). Furthermore, Cushing (1976) stated a paradox: 'Until the stock/recruitment problem is solved, fisheries will fail'.

The only sure knowledge about stock-recruitment relations at present is that large spawning stocks have been known to produce proportionately small recruitment and small spawning stocks are known to have produced proportionately large recruitment. (This might be caused by cannibalism and intraspecific competition.) Often a relatively ambiguous statement is made: 'Recruitment per unit stock decreases when stock increases'. On the other hand, many stocks are remarkably stable. This stability is partly caused by the presence of several year-classes, which can be different in strength. Thus a stock of long-lived species is buffered against large recruitment fluctuations. The shorter the mean life span of the species, the larger fluctuations in stock can be expected (caused *eg* by cannibalism, competition, and environmental factors).

Spawning may be considered (and treated numerically) either as a discontinuous or a semicontinuous process in time (*Fig 22*). There are several justifications for treatment of this recruitment as a continuous process in the biomass-based models: First, the recruitment to the exploitable biomass is continuous because of the growth (and size) diversity in respect to age within a population; second, the spawning time can be several months long, especially if a number of species are included in an ecological group in an ecosystem simulation; and, third, the growth of biomass commences after hatching. However, in most considerations the larvae are accounted for in the species biomass at the age of four to six months; eggs and early (small) larvae are consumed at the same rate as zooplankton in the model.

In fact, the treatment of spawning as a discontinuous process is more 'artificial' than the continuous treatment. The continuous recruitment in biomass-based models is mainly regulated by a time change of the growth coefficient. The change in growth coefficient is also indicative of relative year-class strength variation. When a

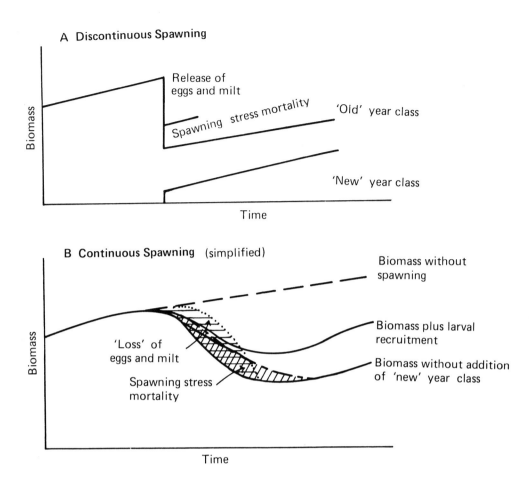

A Discontinuous Spawning

Biomass

Release of
eggs and milt

Spawning stress mortality

'Old' year class

'New' year class

Time

B Continuous Spawning (simplified)

Biomass

Biomass without
spawning

'Loss' of
eggs and milt

Spawning stress
mortality

Biomass plus larval
recruitment

Biomass without addition
of 'new' year class

Time

Fig 22 Schematic presentation of A–discontinuous spawning (which requires the separation of a given species biomass into several age groups), and B–continuous spawning (which can be treated with a given species biomass as a unity).

given biomass is abnormally low, there is a justification of a slightly increased growth coefficient, partly as recruitment compensation (density-dependent) and partly because of possible increased availability of food (density-dependent feeding).

The following paragraphs contain suggestions on the treatment of semi-discontinuous spawning in biomass-based models. The last paragraphs describe the fully continuous treatment of recruitment as applied at present in PROBUB.

(*1*) Determine the period (months) of spawning of the species. If a number of species are grouped into an ecological group, the spawning period of this group might be long indeed (one of the justifications for treatment of spawning as a continuous process).

(*2*) Subtract 1% to 3% of the biomass per month to represent the release of eggs and milt. The amount subtracted depends on the length of spawning period and the biomass distribution with age. The eggs constitute about 10% of the body weight of average spawning; milt is less. An average percentage of 8% of sex products can often be assumed.

(*3*) Increase the growth coefficient (g_0) in the second and third month after first month of spawning. Percentage increase depends, among others, on life span, fecundity, and growth rate of the species. Decrease the growth coefficient gradually in subsequent 5 to 9 months to reach its 'normal' value. This change of growth coefficient simulates the recruitment of the larvae into the biomass of the species.

(*4*) If a species spawns in a relatively restricted area, the released eggs and the early larvae can be added to zooplankton and assumed to be consumed at the same rate in DYNUMES model. If an ichthyoplankton component is considered in the model, about 1/3 of the released egg mass is added to this group, grown fast, and consumed fast.

Another suggested spawning and recruitment subroutine of total biomass is:

(*1*) Introduce the fraction of adult (exploitable) biomass (f_a) as a parameter in the simulation.

(2) Assume that sex products are about 8% of body weight of spawning population (*ie* of the adult biomass f_a).

(*3*) Subtract $0.04 f_a B_e$ of biomass each of two spawning months from the biomass.

(*4*) Apply spawning stress mortality during the two months (Formula 34).

(*5*) Increase growth coefficient for second month and allow this coefficient to return to 'normal' during the following six months.

The recruitment is usually depicted in number-based models as a time-dependent discontinuity, relating it to a discrete spawning period. In the biomass-based model, spawning can be treated as a continuous process. This consideration is more acceptable if we think in terms of size groups rather than age groups, a long spawning period, and consider variations in growth of individuals belonging otherwise to the same age group.

Considering a continuous recruitment to all size groups and assuming there are no exceptionally strong or weak year-classes, recruitment would be proportional to the biomass present, then variations in postlarval recruitment would be depicted in the biomass-based model by the variations of the growth coefficient in the species biomass as long as the species is treated as one unit.

On the other hand, large spawning biomasses are known to produce proportionally smaller year-classes and small spawning biomasses are known to produce proportionally large year-classes. Therefore, the recruitment could be regulated in the biomass-based PROBUB and DYNUMES model, making the growth coefficient inversely proportional to the biomass present, as follows:

$$g_{o,c} = g_o B_{e,i}/B_{i,t-1} \tag{35}$$

where $B_{e,i}$ is the equilibrium or mean biomass of species i. This computation can be carried out in the models in prognostic mode after the determination of the equilibrium biomasses. In the determination of 'equilibrium biomasses', ie finding a unique solution to the set of biomass balance equations, the recruitment problem is eliminated because of the annual adjustment of biomasses.

The factor $B_{e,i}/B_{i,t-1}$ dampens the possible fluctuations of recruitment rather heavily so that the much above or below average recruitment does not appear. It has been found somewhat more acceptable to use the term $\sqrt{B_{e,i}/B_{i,t-1}}$ instead (see Formula 21). It should be noted that in contrast to number-based models, the biomass models are not overly sensitive to errors in recruitment computations, as the biomass is buffered by the presence of many year-classes.

5.6 Migrations and the effects of space-time variation of predator and prey

The migrations of fish can be classified into several categories by their periodic occurrence and by cause. Seasonal migrations are either caused by a desire (instinct) to find proper and abundant food, or by a search for optimum environmental conditions. Life cycle dependent migrations are spawning migrations to traditional spawning grounds, predation avoidance migrations such as outmigration of adults from their own spawning grounds, migration of juveniles either into coastal or offshore regimes, and life cycle dependent feeding migrations (with reference to availability of proper sized food and food composition changes with age). Environment dependent migrations can be affected by seasonal changes and profound environmental anomalies, such as too cold water. The advection by currents and the response of fish to currents belong to this category of migrations.

There are numerous reports and some books on fish migrations; however, the migrations of fish in many areas, such as in the NE Pacific, are poorly known, as very few tagging experiments have been conducted. Often one has to estimate the migrations (especially seasonal migrations and their speeds) from the known summer and winter distributions.

Fish are transported also by currents. Furthermore, they can take advantage of tidal currents during their migrations (Harden Jones et al, 1979). Many environmental factors can affect and 'force' migrations. Koto and Maeda (1965) demonstrated that cold bottom water affects the migrations of flatfish and causes aggregation at the boundaries of cold water areas.

Migrations through the boundaries of models must be treated in an empirical way, adding or subtracting a given conservative amount each time step. These estimates can be based on the knowledge of seasonal migrations. Furthermore, some regions are source regions and some are sink regions of biomass which must be taken into consideration in estimating seasonal migrations through the boundaries as a given biomass which leaves a given region seasonally can be either smaller or larger when it returns. In an advanced state of modeling, the boundary values for smaller, detailed models can be obtained from a large-scale (ocean-wide) gridded model, which would eliminate the guesswork of migrations through boundaries.

Computation procedures for migration through boundaries involve the removal or addition of a predetermined fraction of biomass for months in which migration occurs from one region to another. The return migration in a later month must be adjusted by growth and predation that occurred while the animals were in different regions.

The static conditions which have usually been implied in past studies of ecosystem productivity, as well as in respect to the effects of fishing, will not give quantitative answers when the prey is quasi-stationary and predators migrate, or vice versa. The effects of migration on predator-prey interactions are especially important in gridded models, such as DYNUMES, and is one of the main reasons for the use of such gridded models.

The effects of spatial distribution of different prey items on the composition of food of a predator are schematically shown in *Fig 23* which depicts a vertical section with predator-prey distribution. Not only does the food composition of the predator vary in space, but the predation pressure on the prey items varies as well.

Not all of the biomass participates in a given migration. Thus, before computing migrations in gridded models such as DYNUMES, the portion of migrating biomass must be estimated. The migration speed must be ascertained and separated into u and v components, which must be prescribed or simulated.

The computation of migrations is performed with an 'upcurrent interpolation and advection' method, which is quasi-conservative. However, the conservation of biomass must be checked after each time step. The migration formulation has a stability criterion and might require a shorter time step than the routinely used monthly time step:

$$t_d < \frac{1}{V_{max}} \tag{36}$$

where: t_d is the time step, l is grid length, and V_{max} is the migration speed component.

Migration computation is carried out in two steps. First, the linear gradient of biomass in the 'upcurrent' (upmigration) (UT and VT) is determined:

U positive:
$$UT_{(n,m)} = (B_{n,m} - B_{n,m-1})/l \tag{37a}$$

U negative:
$$UT_{(n,m)} = (B_{n,m} - B_{n,m+1})/l \tag{37b}$$

60

V positive:
$$VT_{(n,m)} = (B_{n,m} - B_{n-1,m})/l \qquad (37c)$$

V negative:
$$VT_{(n,m)} = (B_{n,m} - B_{n+1,m})/l \qquad (37d)$$

In the second step, the gradient is advected to the grid point under consideration.

$$B_{(t,n,m)} = B_{(t-1,n,m)} - (t_d|U_{(t,n,m)}|UT_{(n,m)}) - (t_d|V_{(t,n,m)}|VT_{(n,m)}) \qquad (38)$$

After each time step a smoothing (diffusion) operation is performed, which also simulates the random migration of fish:

$$B_{(n,m)} = \gamma B_{(n,m)}' + \beta (B_{n-1,m} + B_{n+1,m} + B_{n,m-1} + B_{n,m+1}) \qquad (39)$$

where γ and β are smoothing coefficients ($\gamma = 0.8$ to 0.96 and $\beta = (1-\alpha)/4$).

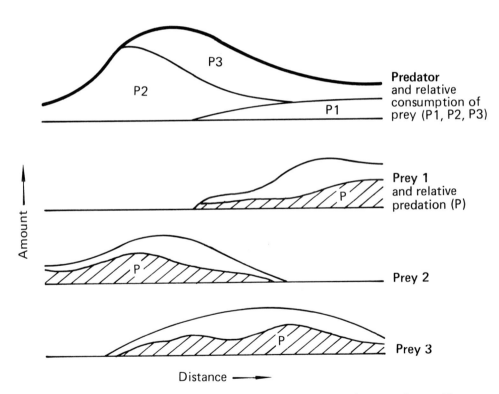

Fig 23 Schematic presentation of the predation by one predator on three prey items with different spatial distributions (presented as a section).

The migrations due to unfavourable environment and/or search for food are carried out by checking surrounding grid points for various prescribed unfavorable–favorable criteria (and/or presence of optimum conditions) and corresponding to the success of this search a portion of the biomass is moved toward optimum conditions:

$$B_{i(n,m)} = B_{i(n,m)} - k_o B_{i(n,m)} \tag{40a}$$

$$B_{i(n\pm1,\,m\pm1)} = B_{i(n\pm1,\,m\pm1)} + k_s B_{i(n,m)} \tag{40b}$$

where: k_o is the fraction of biomass removed from the grid point and k_s is the fraction of this biomass added to a given neighboring point. The size of these fractions depends on the nature of the 'forced migration' and on the number of favorable and unfavorable gridpoints surrounding the grid point under consideration. Consequently the simulation of migrations is in a high degree dependent on local conditions.

6

Formulas and computation procedures of a biomass-based fisheries ecosystem model-PROBUB (Prognostic Bulk Biomass Model) with notes on a gridded model-DYNUMES (Dynamical Numerical Marine Ecosystem Simulation)

Previous chapters describe and quantify some major processes in the marine fish ecosystem which were schematically shown in *Fig 1*. In this chapter the major computation formulas and procedures are summarized according to the computation flow diagram in *Fig 24*.

The formulas refer to the PROBUB (Prognostic Bulk Biomass) model, which uses homogeneous regions ('boxes') (*Fig 6*) as spatial units. Most formulas are also used in the DYNUMES model (Dynamical Numerical Marine Ecosystem Simulation) in the same form, where they are applied at each grid point with coordinates n and m. Additional specific DYNUMES computations (such as migrations) are also given in this section where applicable.

The apex predators are assumed to obtain their food (prey) from the ecosystem without being affected by the fluctuations of abundance of individual prey items. This assumption is based on the high mobility of the apex predators in their search for food and their rather flexible food habits.

The monthly numbers of the mammals and birds in given regions are estimated and converted to biomass, using mean individual weights in the population. As the number of mammal and bird species is large in the NE Pacific, some are grouped into ecological groups using average food composition as the criterion for grouping.

Due to uncertainties in the estimates of the abundance of marine mammals, birds, and other apex predators present in any given region in any given month, it is not necessary to compute their growth and mortality (except in special studies emphasizing the apex predator aspects of ecosystems) as the results of such computations will be embedded in the errors of abundance estimates. Thus, predation by apex predators is computed in the present model as the first step in the computations. The total food requirement of apex predator a in a time step t $(C_{a(t)})$ is:

$$C_{a(t)} = B_{a(t)} q_a t_d \qquad (41)$$

where $B_{a(t)}$ is the biomass of the predator, q_a is the food requirement in terms of

BWD (body weight daily), and t_d is length of time step in days. The consumption of species i by predator a ($C_{i,a(t)}$) is:

$$C_{i,a(t)} = C_{a(t)} p_{i,a} \qquad (42)$$

where $p_{i,a}$ is the fraction of species i in the food of species a.

The consumption by marine mammals serves as a 'forcing function' in the PROBUB model to find the equilibrium biomasses (see Chapter 10). In case the amount of marine mammals is small in a given region (and the fishery is also small in relation to the resources present), a major species, whose biomass can be estimated by other means (such as VPA), should be kept constant as prescribed with initial first guess monthly inputs.

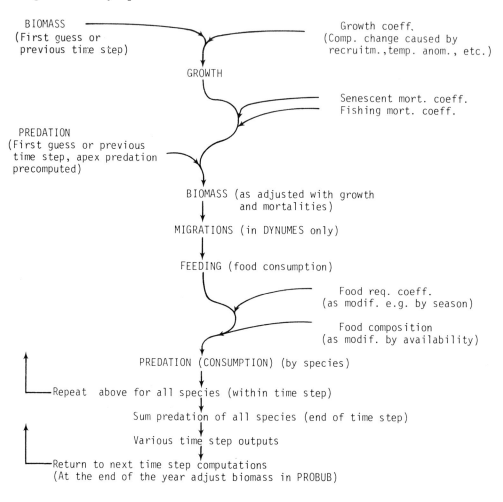

Fig 24 Schematic flow diagram for computations.

The first guess of fish and benthos biomasses are introduced in the PROBUB model and the biomasses iterated to a unique solution with the formulas summarized below. The DYNUMES model is initialized with the equilibrium biomasses obtained from the PROBUB model.

The biomass in time step t ($B_{i(t)}$), and its growth, is computed with a well-known formula similar to compound interest computation:

$$B_{i(t)} = B_{i(t-1)} (2 - e^{-g(t)}) \tag{43}$$

The growth ($G_{i,(t)}$) of the biomass is: $G_{i(t)} = B_{i(t-1)} (1 - e^{-g(t)})$. The growth coefficient $g_{(t)}$ is computed for each time step (and for each grid point in DYNUMES). The time step in the model is sufficiently short so that second order terms can be neglected. The basic growth coefficient (g_o) is obtained from empirical data (see Section 5.1 Chapter 5.) and must correspond to the time step used in the model. The growth coefficient is changed due to factors affecting the growth. In some species it is made a function of time (*ie* seasonal variations in growth).

$$g_{(t)} = g_o + A_g \cos (\alpha t - \kappa_g) \tag{44}$$

where A_g is the half annual magnitude of change of the growth coefficient, α is the phase speed (30° in case of monthly time step) and κ_g is the phase lag, *ie* determining the time when the growth coefficient is at a maximum.

In other 'temperature sensitive' species, the growth is made a function of either surface or bottom temperatures, depending on whether the species is pelagic or demersal in Formula 18 below:

$$g_t = g_o e^{\left(\frac{1}{T_o} - \frac{1}{T}\right)} \tag{18}$$

where T_o is the acclimatization temperature for the species and T is the temperature of the actual month. Furthermore, the growth is dependent on the degree of starvation (see Section 5.1 Chapter 5).

$$g_t = g_o - \left(\frac{S_i}{R_i} g_o\right) \tag{19}$$

The food needed (normal ration) by the biomass of species i in a given location (R_i) and the shortage of food to satisfy the food requirement of the species (S_i) are computed in the model. This starvation effect on the growth of biomass is computed as a 'correction' at the end of the time step when both (S_i) and the ration (R_i) are available from previous computations.

The biomass balance formula (including growth, mortalities, and predation, but excluding fishery), is:

$$B_{i(t)} = B_{i(t-1)}(2-e^{-g(t)}) e^{-m(t)} - C_{i(t-1)} \qquad (45)$$

In the above formula $C_{i(t-1)}$ is the predation of species i in the previous time step and $m_{(t)}$ is the mortality coefficient of senescent, spawning stress, and disease mortalities and is estimated as described in Section 5.4 Chapter 5.

In some species the mortality coefficient is slightly increased at the grid points where severe starvation is computed. Furthermore, a spawning stress mortality is computed for some species during the months of spawning. (Fishing mortality is also bigger during spawning in most species.) It is emphasized that the mortality coefficient $(m_{(t)})$ does not represent the past conventional natural mortality but is about one order of magnitude smaller. The past conventional natural mortality included predation mortality; the latter is, however, being computed in our model separately and in detail. The predation mortality (C) must be taken from the previous time step as its simultaneous computation is not feasible. The taking of a previous time step value of a variable is one of the conventional approaches in finite difference methods in large, time-dependent numerical models.

The computation of consumption (predation mortality) is the most laborious part of the ecosystem simulation. First, we must compute the amount of food eaten by species i with unlimited food availability (or the normal ration of the food) $(R_{i(t)})$:

$$R_{i(t)} = B_{i(t)} r_i t_d + (B_{i(t)} (1-e^{-g(t)}) r_{g,i}) \qquad (46)$$

where r_i is the fraction of body weight of food required daily for maintenance, t_d is the length of time step in days, and $r_{g,i}$ is the food requirement for growth, expressed as the amount of food for growth divided by growth (see further Section 5.3 Chapter 5).

If the supply of all food items for a given species was unlimited, we could compute the consumption of each food item, *eg* the consumption of species j by species i $(C_{j,i})$, from the food requirement (R_i) and the fraction of species j (prey) in the food of species i (predator) $(p_{i,j})$:

$$C_{j,i(t)} = R_{i,t} p_{i,j} \qquad (47)$$

In this case the total consumption (predation) of species i would be:

$$C_{i(t)} = \overset{j}{\Sigma} C_{i,j(t)} \qquad (48)$$

and the starvation would be 0. However, the fraction of species j in the food of species i $(p_{i,j})$ can vary in space and time, depending on the availability of species j (density dependent feeding). Thus the food composition of the species at each time step (and location) must be computed as described in Section 5.3. Furthermore, this computation will also yield the quantitative parameters of partial starvation.

The yield (Y) is computed with a prescribed fishing mortality coefficient ϕ_i. It should be noted that all the instantaneous coefficients (growth, mortality, fishery) are different from the corresponding conventional coefficients for number-based models which use an annual time step. Thus all these coefficients have to be computed on a biomass basis and for the time step used in the model (also considering the amount of biomass present and the catch statistics) (see further Chapter 14).

$$Y_{i(t)} = B_{i(t)} - B_{i(t)} e^{-\phi_{i(t)}} \tag{49}$$

The yield is subtracted from the biomass at the end of each time step. In the DYNUMES model the fishing intensity coefficient can also vary in space.

Proper computation of migrations can be effected only in gridded models such as DYNUMES. As in some cases only a portion of the biomass undertakes the migration (*eg* seasonal depth migrations of flatfish, spawning migrations), the portion of biomass involved in the migration must be estimated and separated quantitatively from the rest before computing migrations.

The U and V components of migration speeds are determined from empirical data. An 'upcurrent interpolation and direct advection' quantitative migration computation has been found mass conserving and applicable in our model. This computation is carried out in two steps, first the linear gradient of biomass in the 'upcurrent' is determined:

U positive:
$$UT_{(n,m)} = (B_{n,m} - B_{n,m-1})/l \tag{37a}$$

U negative:
$$UT_{(n,m)} = (B_{n,m} - B_{n,m+1})/l \tag{37b}$$

V positive:
$$VT_{(n,m)} = (B_{n,m} - B_{n-1,m})/l \tag{37c}$$

V negative:
$$VT_{(n,m)} = (B_{n,m} - B_{n+1,m})/l \tag{37d}$$

Thereafter the gradient is advected to the grid point under consideration:

$$B_{(t,n,m)} = B_{(t-1,n,m)} - (t_d|U_{(t,n,m)}|UT_{(n,m)}) - (t_d|V_{(t,n,m)}|VT_{(n,m)}) \tag{38}$$

If a species is transported by current, the same method is used for computation of advection, except the U and V current speed components are used. This calculation (38) has a stability criterion that the time step must be smaller than $1/V_{max}$.

After migration computations a smoothing (diffusion operation) is performed, which represents random movements of fish, which is assumed to occur also on fishing and feeding grounds and after spawning. This 'smoothing operation' is applied several times in the computations:

$$B_{(n,m)} = \gamma B_{(n,m)} + \beta (B_{n-1,m} + B_{n+1,m} + B_{n,m-1} + B_{n,m+1}) \qquad (39)$$

γ is a diffusion coefficient (0·80 to 0·96) and β is $(1-\alpha)/4$.

Fish migrate in search of food and can be forced to alter paths if environmental conditions get unfavorable, *eg* too cold or too warm. These migrations are treated in the DYNUMES model in the following manner: Each species is assigned upper and lower limits of optimum temperature and lower food concentration (density) limits for the most important food items. Tests are made at each grid point and time step of these criteria. If the criteria are exceeded, the gradients of surrounding points are tested and a portion of the biomass at unfavorable grid points is moved towards more favorable conditions:

$$B_{i(n,m)} = B_{i(n,m)} - k_o B_{i(n,m)} \qquad (40a)$$

$$B_{i(n\pm1,m\pm1)} = B_{i(n\pm1,m\pm1)} + k_s B_{i(n,m)} \qquad (40b)$$

The coefficient k_o depends on the amount values exceed the prescribed criterion and on the number of computational passes (about 0·03 in a two-pass operation). The coefficient k_s depends in addition on the number of favorable surrounding points (two-pass value is about 0·008 to 0·03).

68

7

Simulation of plankton and benthos, interactions of fish biota with them, and the total carrying capacity of ocean regions

This chapter reviews briefly the quality and quantity of empirical data on plankton and benthos in the NE Pacific. Formulas and procedures for simulation of the standing stocks of plankton and benthos are given, together with a few computed examples.

The role of plankton and benthos in the fish ecosystem and their production and consumption are briefly summarized and illustrated with some computed numerical values. This chapter also describes how the plankton and benthos production determine the carrying capacity of any given ocean region.

Phytoplankton production is the main constituent of basic organic matter in the sea and thus might determine on a large space scale the quantitative level of a marine ecosystem (*ie* large-scale total carrying capacity). The direct quantitative utilization of phytoplankton as food by other biota in the marine ecosystem is variable in space and time. Phytoplankton also produces the bulk of the organic matter for the remineralization process, for burial in sediments, and for use as detrital food by benthos.

The most important direct and basic food source (and 'food buffer') for marine biota is, however, the zooplankton (which includes euphausiids and epibenthic crustaceans). The quantitative knowledge on zooplankton, and especially about its production, is deficient in many areas and seasons in the NE Pacific Ocean. The values for phytoplankton and zooplankton standing crop are relatively few in this region and vary more than one order of magnitude as reported by various authors. Thus, it is difficult to obtain a reliable analysis of the seasonal variation of phyto- and zooplankton in space and time. The variability of phyto- and zooplankton production estimates in the Bering Sea is demonstrated below with a few examples. The estimates of annual primary production in the eastern Bering Sea range from in excess of 300g C/m² (Alexander, 1978) to lower values such as 150g C/m² 'inshore' (Motoda and Minoda, 1974) and 35g C/m² 'offshore' (Smetanin, 1956; Heinrich, 1962). Motoda and Minoda (1974) reported that the zooplankton standing stock (mainly copepods) in the Bering Sea during the summer varies from about 20g to 67g biomass/m². Accurate quantitative data on abundant euphausiids is almost

69

entirely absent. For copepod production, low values (115g to 135g biomass/m²/year) have been reported (Heinrich, 1962).

Considering the deficient quantitative knowledge on phytoplankton and zooplankton production, it becomes obvious that the results of quantitative trophodynamic ecosystem simulation would be unreliable, if one started the computations with basic organic and/or zooplankton production, as has been customarily done in many earlier attempts to compute marine production in 'food chain' type models. Furthermore, the pathways of plankton production through the food chain are extremely variable in space and time and the 'transfer rates' are quantitatively nearly unknown (see Chapter 3).

One of the most important factors in respect to the zooplankton availability to fish is its patchiness, about which our knowledge is at best qualitative. The best approach to use zooplankton rationally in ecosystem computations is to simulate its standing stock based on the best available empirical data and to limit its consumption in space and time in the same manner as that of other ecological groups (ie allowing a given amount of standing stock to be consumed in unit time).

The estimates of how much of the zooplankton standing crop can be consumed in any given time step (month) depend on the estimated turnover rate of the zoo-plankton which can vary seasonally and latitudinally. Similarly, the phytoplankton consumption by the biota may be computed in each time step, including the phytoplankton consumption by zooplankton and the consumption of both plankton components as detritus by benthos.

The standing stocks of phyto- and zooplankton are simulated with a harmonic formula, which can be tuned to reproduce the known and/or estimated seasonal cycles:

$$P_t = P_0 + A_1 \cos{(\alpha_1 t - \kappa_1)} + A_2 \cos{(\alpha_2 t - \kappa_2)} \tag{50}$$

where P_t is the standing stock (separate computations for phyto- and zooplankton), P_0 is the annual mean of this standing stock, A_1 and A_2 are the first and second harmonic constants corresponding approximately to the half-range of the main and secondary annual range of the variation of the standing stock), α_1 and α_2 are the phase speeds (30° and 60°, respectively, if a monthly time step is used), t is time (in months) and κ_1 and κ_2 are the phase lags which determine the times of the primary and secondary 'peaks' of standing stocks. An example of simulated phyto- and zooplankton standing stocks in two regions in the NE Pacific are given in *Fig 25*. In the gridded DYNUMES model the spatial distribution of plankton standing crop can be simulated at each grid point through spatial variation of P_0, A_1, A_2 and phase lags (α_1 and α_2).

The data on benthos and its production are still more deficient for the Bering Sea than are the data on plankton. Almost nothing is known on the annual production of different components of benthos. The quantitative data on benthos (Alton, 1973)

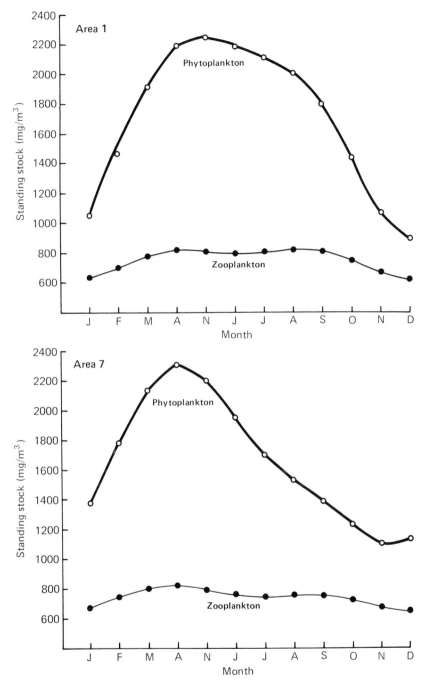

Fig 25 Annual changes of phytoplankton and zooplankton standing stocks in subregions 1 and 7.

71

can be summarized as follows: The total benthos biomass ranges from 55g/m² to 905g/m²; the average value for the north central part of the Bering Sea is 170g/m², and the overall mean is 100g/m². The highest standing stocks of benthos occur in depths between 50m and 150m. Some general information on the dependence of benthos communities and their biomasses on type of the bottom, depth, distance from the coast, and bottom temperature is available in literature from many areas in the North Atlantic. This information has been used for simulation of benthos, especially in DYNUMES, where different standing stocks can be simulated at each grid point, using the determining conditions there.

For the purpose of the fish ecosystem simulation with the PROBUB model where benthos is the second 'production buffer', it is generally sufficient to treat benthos as three ecological groups – predatory benthos, infauna, and epifauna. Each of these groups is given a species composition, growth rate, and other input parameters similar to the fish biota components, and the equilibrium standing stocks of these benthic ecological groups are simulated (see Chapter 8). The simulated values are restrained (and tuned), if necessary, based on empirical knowledge on average standing stocks of these benthos components.

The plankton and benthos standing stocks (and production) participate in the determination of the upper limits of the total biomass in a given region (*ie* the total carrying capacity); and therefore have been designated as 'production buffers'. However, it has to be borne in mind that they are not the only factors determining the total carrying capacity of any marine ecosystem, as many other environmental and ecosystem conditions participate in the determination of the carrying capacity. The zooplankton and benthos control the productivity only on large space and time scales and should be used as such in the fish ecosystem models.

Fish and other mobile marine organisms seem to be in constant search for food; in contrast, most of the benthos and plankton have limited or no mobility at all. It is generally assumed that there is a migration by fish into high food density areas and when the food is grazed down in these areas the grazers move into other areas, leaving former grazed down areas to recover. Fish locate high food concentrations by encounter. Therefore, there exists no generally valid quantitative relation in small- and medium-space scale between the production of fish and its food.

Examples of 'saturation equilibrium biomasses' or total carrying capacities of some ocean subregions (coastal, continental slope, and offshore regions) near Kodiak Island in the Gulf of Alaska are given as examples in *Tables 3* and *4*. The utilization of phytoplankton in this area is about 20% of the estimated production. Zooplankton utilization over the shelf is very high, as compared to estimated production. This can be interpreted to mean that there might be horizontal (and vertical) advection of zooplankton to the continental shelf with the seasonally prevailing upwelling circulation in this region. In a detailed study it was found that the estimated plankton and benthos productions could sustain the computed saturation biomasses in these regions (see also *Table 4*).

Table 3
Mean biomass (B) (tonnes/km²), annual consumption (C) (tonnes/km²), and turnover (T) in coastal, continental slope, and offshore subregions in Kodiak area from PROBUB model.

Species and/or ecological group	Coastal subregion			Slope subregion			Offshore subregion		
	B	C	T	B	C	T	B	C	T
Herring	7·32	7·71	1·05	3·03	2·86	0·94	1·71	1·60	0·94
Other pelagic fish	12·47	13·47	1·08	12·62	11·55	0·92	7·32	6·86	0·94
Squids	2·54	2·89	1·14	2·31	2·21	0·96	1·52	1·43	0·94
Salmon	0·41	0·15	0·37	0·34	0·11	0·32	0·37	0·16	0·43
Rockfishes	2·45	2·24	0·91	1·57	1·22	0·78	0·44	0·31	0·70
Gadids	6·74	6·92	1·03	4·52	4·31	0·95	1·32	1·08	0·82
Flatfishes	3·10	2·72	0·88	1·87	1·50	0·80	0·42	0·29	0·69
Other demersal fish	3·84	3·74	0·97	3·30	2·82	0·85	0·65	0·48	0·74
Crustaceans (commercial)	7·15	8·66	1·21	3·62	4·02	1·11	1·34	1·52	1·13
Benthos ('fish food' benthos)	36·85	32·20	0·87	19·42	16·39	0·84	3·39	2·57	0·76
Total finfish	36·33			27·25			12·23		

Numerical examples in *Tables 3* and *4* demonstrate the importance of benthos in determining the carrying capacity of fish in any ocean region and demonstrate why the standing stock of fish is in general higher over the continental shelf than over the deep water. Furthermore, additional studies with the PROBUB model indicate that the carrying capacity of any ocean region is also largely dependent on the composition of fish biota and their principal food.

An urgent need exists for quantitative studies of benthos standing stock and production, as well as for determination of the standing stocks of euphausiids and their production. These studies should, however, be directed by the ecosystem simulation needs and emphasize aspects which can be generalized in space and time.

Finally, it should be pointed out that the fishery can affect the large-scale carrying capacity of any given region in several ways. One way is by lowering the carrying capacity in the same manner as that by marine mammals (see Chapter 12, *Fig 44*). Another way is by increasing the carrying capacity *eg* by removing larger cannibalistic fish, relieving predation pressure on juveniles, and thereby allowing the biomass to increase (see Chapter 11, *Fig 42*).

Table 4
Estimated plankton productions, standing stock, and their consumption in coastal, continental slope, and offshore subregions in Kodiak area using PROBUB model.

Subject	Coastal subregion	Slope subregion	Offshore subregion
Annual mean phytoplankton production and mean	1500	1350	1000
standing crop, tonnes/km^2	200	180	135
Annual mean zooplankton production and mean	225	180	200
standing stock tonnes/km^2	45	36	40
Annual phytoplankton consumption by nekton tonnes/km^2	20	14	6
Annual phytoplankton consumption by zooplankton tonnes/km^2	241	194	220
Annual zooplankton consumption by nekton tonnes/km^2	129	123	58
Annual consumption of detritus by benthos tonnes/km^2	133	70	12

8

Equilibrium biomasses and basic input data

Equilibrium biomass is defined as the biomass of a given species (or group of species) which can be sustained in a defined region, assuming that the growth of the biomass equals the sum of mortalities. Consequently the equilibrium biomass would be the same in a defined month (January) from one year to another. This requires that growth as well as mortalities (especially predation) remains at the same level each year (although seasonal fluctuations occur). The equilibrium biomass concept may be considered an unnatural one, but it is required as a 'standard' (basis) for the assessment of long-term mean resource (biomass) levels and their internal quantitative relations. The equilibrium biomasses are determined as a unique solution (a singular point where the 'balance' changes from positive to negative) to the set of basic ecosystem biomass balance equations:

$$B_{e,(const.)} = B_{(adj.)}(2 - e^{-(g-Z)}) \tag{51}$$

In other words: If the biomasses of all species in the ecosystem do not change over a year (*ie* the previous January biomass is the same as subsequent January biomass), then we can say that the biomasses are at the equilibrium point. This implies that the growth of the biomass equals its removal by mortalities (especially by predation) $(g-Z=0)$. If we want to achieve this equilibrium, we can change either the growth rate or mortality rate, or the biomass level itself must be adjusted. However, the growth rate is determined by empirical data (and other factors such as temperature) and is assumed in the equilibrium case to be the same from one year to another (although seasonal changes can occur). Fishing and other mortality rates are also assumed to remain (in the equilibrium case) the same from one year to another. The predation mortality (consumption) must then balance (together with other mortalities which remain unchanged) the growth rate. This balancing can be achieved if the levels of the biomasses are adjusted iteratively in the simulation model at the end of each year so that at the end of the iterations the biomass of one January is the same as in the next January. This adjustment can be done by finding a simultaneous unique solution to the biomass equations of all species (or groups of species) in the ecosystem. This unique solution exists only as a 'point' either if one of the major biomasses and the consumption by that biomass is predetermined (assumed to be known

75

and fixed), or there exists a predetermined predation by apex predators. In this case an iterative procedure can be applied to adjust the biomasses of other species once after each year's computation:

$$B_{i,t_{12},0} = B_{i,t_{12},a} + \frac{(B_{i,b} - B_{i,a})}{k} \qquad (52)$$

where $B_{i,t_{12},0}$ is the new (adjusted) biomass for December, $B_{i,t_{12},a}$ is the previous December biomass, $B_{i,b}$ is the biomass of previous January (computed as next step from $B_{i,t_{12},a}$), $B_{i,a}$ is the computed biomass in January one year later and k is an iteration constant (changed during iteration from 3·5 to 10, depending on the state of convergence). Forty years or more of computation is usually needed before the solution converges to a unique (equilibrium) solution. The speed of convergence is dependent among other factors on how close equilibrium values were to the initial guess biomasses at the start of the computation.

We can define three basic equilibrium biomasses which are useful for determination of our numerical assessment error limits: **Minimum equilibrium biomasses,** computed using lowest plausible food requirements and highest plausible growth rates; **maximum or saturation equilibrium biomasses,** computed using lowest plausible growth rates and highest plausible food requirements; and, **mean equilibrium biomasses** obtained using plausible mean values for both requirements.

Errors in the initial (first guess) input of biomasses do not affect the results but affect the time (number of iterations) required for convergence. However, to obtain a unique solution, part of the predation mortality must be known (fixed). This can be done by keeping the biomass of a major species whose magnitude is well known constant, or by predescribing marine mammal biomasses and precomputing the predation by them.

There must be a trophic relation between the biomasses in the system if the system is a **unit** ecosystem (*Fig 26*; in (A) there is one system whereas in (B) there are two independent systems).

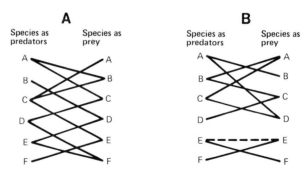

Fig 26   Trophic interactions in ecosystem as determinant of the system dependence. A–one system; B–two systems.

As mentioned above, the food requirements (rations) and growth rates are determining factors of biomass levels in the ecosystem; food composition determines the 'bond' of the species in the system and relative predation of different species.

If the biomass is in equilibrium with growth and fishing and other mortalities:

$$B(1-e^{-g})=B-Be^{-\left(F+m+\frac{C}{B}\right)} \tag{53}$$

where C is predation, F is fishing mortality, and m is other mortalities (senescent, disease, and spawning stress mortalities).

Then:

$$Be^{-g}=Be^{-\left(F+m+\frac{C}{B}\right)} \tag{54}$$

if $F=0$ then $g=\dfrac{C}{B}+m$

However, if g remains constant and F increases, C/B must decrease (which is only partly valid in density dependent feeding). However, if F increases, g must also increase which leads to the rejuvenation of populations. Furthermore, if F increases, m must logically decrease due to a decrease in senescent and spawning stress mortalities. (Note, however, that during spawning season both fishing and spawning stress mortalities might be high.) Consequently, an equilibrium biomass can exist only if the fishery on all species remains constant. This leads to the following basic dogma which is utilized in the development of new approaches in fisheries management in Chapters 15 and 16 – *ie* **if the fishery of any species changes considerably in space and time, the whole fish ecosystem must change** (or at least the biomasses of species which are using the changing target species as food, or which serve as food items to the target species, must change).

A word of caution is necessary when determining equilibrium conditions with the present fishery in areas of relatively high fishing intensity. If the fishery on any species is intense, the biomass of this species might not be in equilibrium with the fishery, but might be in a declining state. In this case, the model would overestimate a biomass, unless the proportion of this species in the food of other species is lowered to compensate for high fishing effects. Another method would be to determine the present state of the biomass of heavily fished species with virtual population analysis and prescribe the monthly biomasses.

The above problem leads to another limitation, especially in low fishing intensity regions. Although we can determine the equilibrium biomasses with mean data, we might not be able to determine the actual state of the biomass because of the lack of data and because of the natural fluctuations of biomasses. In this case we must try to obtain recent, reliable data from major surveys which might determine the state of at least one major species or use other possible indices of the state of some more abundant stocks.

The equilibrium biomasses can also be determined in 'natural conditions' (*ie* without the intervention by man) with the PROBUB model, in which the fishery is excluded and growth rates and spawning stress mortalities are adjusted correspondingly (see Chapter 9).

The first **input of the model** is the definition of the region (area). The PROBUB model uses defined regions ('boxes') as basic geographic units (see *Fig 6*); these regions are assumed to be homogeneous in respect to distribution of biomasses and environmental conditions. The gridded DYNUMES model uses a computational grid over a defined region, thus allowing spatial resolution of distributions as well as space-dependent processes. The computation grid should be designed preferably on an equal area projection (*eg* polar stereographic) and should be a subset of an internationally used meteorological grid, which facilitates input of synoptic environmental data. The boundaries of the computation area should preferably coincide with the boundaries of natural (or faunal) regions, alleviating errors in the estimation of migrations through boundaries. A computation area with a small grid and several open boundaries, such as the one presented in *Fig 27*, often requires the specification of seasonal migrations of many species through the boundaries, for which good data are usually not available and subjective estimates must be made.

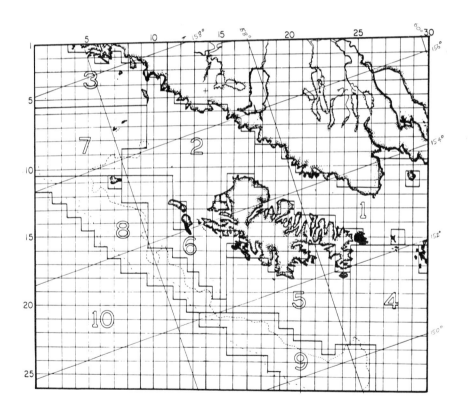

Fig 27 Computational grid with subregions for Kodiak area in Gulf of Alaska.

78

The grid mesh length (grid size) selection is often dependent on the size of the region selected for the model, as well as on the core size of the computer to be used. In past work we have used mesh lengths from 18·5km (Kodiak region in *Fig 27*) with 780 grid points to 190km (North Pacific salmon ecosystem model). It is also advantageous to divide the computation regions in the DYNUMES model into subregions. A digitized index field is prepared and input into the program, in which the grid points over land are designated 0 and over the ocean 1 or the actual subregion number at these gridpoints falling into the designated subregion (see *Fig 27*). These subregion designators allow easy analyses of some subregion dependent distributions.

At each grid point the digitized depth (in meters) and monthly long-term mean surface and bottom temperatures are also required. Temperature anomalies are introduced by changing the digitized long-term monthly mean temperatures at desired grid points and/or subregions.

The subregions of the PROBUB model 80-1 are shown in *Fig 6*. The subregions coincide with statistical and fisheries management areas in the NE Pacific. Each statistical/management area along the coast is divided into two subregions: a shallow subregion from the coast to the 500m isobath, and a deep subregion from the 500m isobath to 200 nautical miles offshore. The latter boundary is obviously rather arbitrary for many biological distributions. The areas of the subregions (in km²) must be prescribed with inputs.

The inputs of biological data are considerably dependent on the region under consideration and on the availability of proper data. Thus, only a few general notes on this subject are given below, with some examples given in tables. Seasonal numbers of apex predators (marine mammals, birds, and sharks) are obtained for each subregion from available data and interpolated to months (no complete monthly distributions are usually available), the numbers are converted to weights, and the resulting biomasses are distributed for grid points in the subregions in the DYNUMES model.

The inputs of first guess field of fish species/ecological groups (*eg* for January) for the DYNUMES model originate from preanalysis with the PROBUB model (equilibrium biomasses) or from extensive exploratory surveys. The data from exploratory surveys must be corrected to total biomasses, including juveniles, using catchability coefficients. The distribution of biomasses between juvenile stages and the exploitable part of the biomass is obtained from another auxiliary model, BIODIS (see Chapter 4).

The PROBUB model requires initial guess input of fish biomasses by species or groups of species in terms of biomass per km². Examples of species and/or ecological groups used in PROBUB 80-1 are listed in *Table 5*. Only a few species, such as herring and pollock, are considered as single species; most are grouped into ecological groups whereby the feeding habits were used as main criterion (observe the groupings of 'largemouth' and 'smallmouth' flatfishes in *Table 5*). Notes on the relative abundance of biomasses of species (*ie* the biomass ratios) in some ecological groups are also indicated for flatfishes in this table. The species/ecological groups are usually referred to in the

Table 5

Species and ecological groups (including numerical relations within some groups) in the Bering Sea and Gulf of Alaska.

---

1 – 4 Species under special study (by age groups)

Demersal (L – 'largemouth', S – 'smallmouth')

5 – Greenland halibut (turbot), Pacific halibut (L) (3·5:1 in Bering Sea)
6 – Flathead sole, arrowtooth flounder (L) (4:1 in Bering Sea; 1:2 in Gulf of Alaska)
7 – Yellowfin sole (until Vancouver Island), rock sole, Alaska plaice (S) (9:1·5:1 in Bering Sea; 2:8:0·5 in Gulf of Alaska)
8 – Other flatfishes (S) longhead dab, Dover sole, rex sole (last two in Gulf of Alaska)
9 – Cottids and others (*eg* elasmobranchs and other noncommercial demersal fish)

Semi-demersal

10 – Pacific cod, saffron cod (saffron cod, polar cod in northern part of Bering Sea)
11 – Sablefish (or black cod, single species)
12 – Pollock (single species)
13 – Pacific ocean perch (and other *Sebastes* and *Sebastolobus* spp.)

Pelagic

14 – Herring (Pacific herring, single species)
15 – Capelin, other smelts, sand lance, and other noncommercial pelagic fish
16 – Atka mackerel and other greenlings, macrourids.
17 – Salmon (5 species, temporary presence)
18 – Squids (mainly gonatids)

Crustaceans

19 – Crabs (King and Tanner crabs and noncommercial species)
20 – Shrimp (several commercial and noncommercial species)

Benthos

21 – Predatory benthos (starfishes and other mobile predatory benthos)
22 – Infauna (annelids and other burrowing forms)
23 – Epifauna (bivalves, benthic crustaceans)

Plankton

24 – Phytoplankton
25 – Copepods
26 – Euphausiids (including sagittas)
27 – Ichthyoplankton (temporary)

---

latter tables by a single representative species of the group. The species group space for 'species 1 to 4' is reserved for special studies of single species which are taken out from a given ecological group and divided into four age (size) groups, when special studies are conducted.

The phyto- and zooplankton biomasses are simulated in the model and tuned to available empirical data as described in Chapter 7. The components of benthos biomass is simulated in the DYNUMES model on the basis of empirical knowledge of standing stocks of the benthos component in respect to depth, distance from the coast, and type of bottom. In the PROBUB model these benthos biomasses are computed in the same manner as fish biomasses.

The migration speeds of individual species/ecological groups must be digitized at the grid points as u and v components in km/day in the DYNUMES model with reference to given months when these migrations occur (see Section 5.6 Chapter 5). These speeds are deduced from empirical knowledge about the seasonal occurrence and migrations including feeding and spawning migrations. The migrations due to unfavorable temperature and scarcity of food are simulated within the model.

Many species-specific coefficients are needed in the simulation models, the annual mean values of which must be prescribed by input. The biomass growth coefficient varies with age, and its derivation from empirical data requires the knowledge of the distribution of biomass with age, which is computed in an auxiliary model (see Chapter 4). The mean growth coefficient must correspond to the computation time step used in the model. The real growth coefficient is computed for each time step and for each area/grid point in the model. *Table 6* gives examples of the computed growth coefficients for two species and seven computation areas for two different months. *Table 7* gives mean growth and mortality coefficients and food requirement coefficients used in PROBUB model 80-1.

The dominant component of natural mortality, *ie* the predation mortality, is computed in detail within the model itself. The mortalities from old age and diseases are small in exploited populations as compared to predation and fishing mortalities. These mortalities can be estimated subjectively or by the use of the above-mentioned auxiliary model (BIODIS, Chapter 4) which gives the total mortality in different age

Table 6

Computed monthly growth coefficients of pollock and flathead sole in February and August in seven subregions of the Bering Sea and Gulf of Alaska. (See *Fig 6*.)

Month and Species	Subregions						
	1	2	3	4	5	6	7
Pollock							
February	0·0715	0·0755	0·0724	0·0840	0·0763	0·0780	0·0768
August	0·0929	0·0903	0·0917	0·0870	0·0829	0·0831	0·0818
Flathead sole							
February	0·0491	0·0519	0·0462	0·0540	0·0488	0·0536	0·0487
August	0·0627	0·0621	0·0580	0·0564	0·0529	0·0571	0·0526

Table 7
Mean growth and mortality coefficients and food requirements for the PROBUB model 80-1.

Species/ecological groups	Species number	Growth coeff. inst. (monthly)[1]	Mort. coeff. inst. (monthly)[2]	Fishing mort. coeff. inst. (monthly)[3]	Food requirements for maintenance % BWD[4]	for growth
Greenland halibut, Pacific halibut	5	0·054	0·007	0·009	0·50	1·32
Flathead sole, arrowtooth flounder	6	0·055	0·008	0·0089	0·50	1·32
Yellowfin sole, rock sole, Alaska plaice	7	0·058	0·009	0·025	0·50	1·32
Other flatfishes	8	0·063	0·008	0·005	0·51	1·32
Cottids, elasmobranchs and other demersal species	9	0·068	0·009	—	0·54	1·67
Pacific cod, saffron cod	10	0·073	0·007	0·0062	0·54	1·63
Sablefish	11	0·072	0·006	0·019	0·54	1·67
Pollock	12	0·080	0·008	0·016	0·57	1·63
Pacific ocean perch + other rockfish	13	0·081	0·007	0·0081	0·51	1·50
Herring	14	0·092	0·006	0·0022	0·54	1·58
Capelin, sand lance	15	0·094	0·008	—	0·54	1·76
Atka mackerel	16	0·072	0·007	0·0022	0·54	1·50
Salmon	17	0·15	0·003	0·04	0·57	1·94
Squid	18	0·18	0·02	0·0015	0·51	1·94
Crab	19	0·06	0·007	0·0022	0·43	1·26
Shrimp	20	0·078	0·008	0·0002	0·43	1·23
Predatory benthos	21	0·070	0·008	—	0·43	1·58
Infauna	22	0·14	0·01	—	0·34	1·76
Epifauna	23	0·10	0·009	—	0·37	1·94

[1]Annual mean, changes with temperature and starvation.
[2]Annual mean, changes with subregions, temperature and severe starvation.
[3]The fishing mortality coefficient given here is for relative guidance only. Each subregion has been assigned proper monthly fishing mortality coefficient.
[4]Annual mean, changes with remperature.

groups, from which predation and fishing mortalities are subtracted, leaving the remainder as old age and disease mortality. Starvation mortality is estimated in the model during computations by increasing the old age mortality slightly in proportion to the degree of starvation at the times and grid points where food availability computations so indicate.

The food requirement coefficient is divided into two parts in the present model: food requirement for growth and for maintenance. The values of the coefficients vary from species to species, depending on activity, growth rates, and normal environmental temperature (*re* metabolism). The food requirement for growth is given as the ratio of growth/food required – and this ratio varies between 1:1·23 and 1:1·94. The food requirement for maintenance is given as a percentage of body weight daily (varying between 0·34% and 0·57%).

Seasonal or annual mean composition of food (in percentages) must be prescribed for each species/ecological group, using available data and considering size-dependent feeding and change of food composition with the age (size) of the species. The food composition tables are used as guidance in the model or as vulnerability to predation indices; the actual composition of food can vary from grid point to grid point, depending on relative abundance of food items. The trophic interspecies interactions in the model are largely dependent on the relative species composition of the food of all species. An example of food composition changes in space and time for flathead sole is shown in *Table 8*. The food composition changes in most flatfishes are relatively small. Studies of these food composition changes and their causes and consequences will be one of the major subjects of the ecosystem simulations studies.

Table 8
Examples of food composition changes in space and time in flathead sole in shallow subregions.

| Food item (group) | Initial | Food composition (in %) | | | |
| | | Subregion 1 | | Subregion 6 | |
		March	September	March	September
Infauna	18	14·3	13·0	18·3	18·4
Epifauna	42	43·9	44·5	42·8	43·0
Euphausiids	9	9·4	9·5	9·2	9·2
Cottids and other demersal species	9	9·4	9·5	9·2	9·2
Cod	4	4·2	3·2	3·0	2·8
Crab	3	3·1	3·2	3·1	3·1
Shrimp	3	3·1	2·3	2·2	2·1
Pollock	8	8·4	8·5	8·2	8·2
Other flatfish	2	2·1	2·1	2·0	2·0
Capelin and other pelagic species	1	1·0	1·1	1·0	1·0
Rockfish	1	1·0	1·1	1·0	1·0
Starvation		0·1	3·0	0	0

An extensive ecosystem simulation requires many additional species- and location-specific data, such as: indices whether the species is demersal, semidemersal, or pelagic; its temperature tolerance limits; schooling habit index; spawning times, *etc*. A large ecosystem simulation model eventually becomes also a depository of all species and location-specific information and/or is connected to a large-scale fisheries information system.

9

Equilibrium biomasses in the eastern Bering Sea at present and Bering Sea biomasses in natural state

Several methods are in use for assessment of marine fishery resources, such as direct surveys, virtual population analyses using catch and age composition data, *etc*. None of these methods are fully sufficient *per se* for resource evaluation, especially not in data sparse areas such as the NE Pacific. Further none of them give either the biomasses of all species and/or ecological groups present, or the productivity of their biomasses per unit time and/or area. The evaluation of total production, starting with primary production, has not been successful because (*1*) the pathways of organic matter transfer are greatly variable in space and time and unknown, and (*2*) the concept of distant trophic levels has been abandoned as an unrealistic oversimplification for quantitative resource assessment.

To overcome the difficulties of resource assessment, described above, the PROBUB model was originally designed for the assessment of the fishery resources and the total marine biota in the NE Pacific. The model was applied to selected regions which were designed so as to coincide with fisheries statistical areas. The model inputs have been documented at some length in various NWAFC technical reports and a few examples of input were given in Chapter 8. To obtain a quantitative picture of the marine ecosystem in the natural state, the model was also run without the effects of the fishery.

The species composition of ecological groups used for the evaluation of fishery resources in the Bering Sea is given in *Table 5*. The maximum and minimum equilibrium biomasses of species and ecological groups in the eastern Bering Sea, as computed with PROBUB model, are given in *Table 9*. (The minimum equilibrium biomasses were estimated with an earlier version of PROBUB which differed slightly from PROBUB run 80-1 only in the food substitution computation.) The results are also given in *Fig 28* where the biomasses are divided into three groups: pelagic, semidemersal, and demersal. The mean exploitable biomasses are also given in *Table 9*.

Examples of the validation of computed equilibrium biomasses with trawling survey results are given in Chapter 13. The PROBUB model does not have much spatial resolution. Therefore, the gridded model DYNUMES was applied to the Bering Sea.

Table 9

Maximum equilibrium biomasses of species and ecological groups in the eastern Bering Sea (in 1,000 tonnes). (Estimated minimum equilibrium biomasses and mean exploitable biomasses are given for comparison.)

Species/ecological group designation	Maximum equilibrium biomass	Estimated minimum equilibrium biomass	Mean exploitable biomass
Halibut	585	400	220
Flathead sole	875	650	380
Yellowfin sole	1,660	1,100	510
Other flatfish	1,160	850	245
Cottids	4,438	4,000	—
Cod	1,468	1,000	745
Sablefish	183	120	51
Pollock	15,165	8,000	6,450
Rockfish	1,825	1,000	485
Herring	2,327	1,500	590
Capelin	5,149	3,500	(1,000)[2]
Mackerel	1,438	1,100	520
Salmon	(73)	(50)	—
Squid	2,310	1,200	(500)[2]
Crab	1,225	800	(300)[2]
Shrimp	1,792	900	(600)[2]
Predatory benthos	818	700	—
Infauna	24,219	20,000	—
Epifauna	20,947	15,000	—
Zooplankton	58,430[1]	35,000	—

[1]$500 mg/m^3$; 100m depth.
[2]Includes species which are not exploited at present.

The latter model showed that the main biomass of the most abundant species, pollock, is over deep water where the species exists in a dispersed state, thus no profitable fishery is possible on this species in this area.

Another validation check of the resources can be effected by comparing the computed maximum and minimum equilibrium biomasses (*Table 9*). The greatest differences in maximum and minimum biomasses occur in pelagic and semidemersal species, especially in pollock, capelin, and other noncommercial pelagic species, as well as in squids. Squids have a short life span and their biomass can vary considerably from year to year. Furthermore, adult squids feed predominantly on other pelagic species (including pelagic juveniles of demersal species) (Akimushkin, 1963). Thus the abundance of squids might exercise considerable influence on other biomasses.

The annual turnover rates (predation and other mortalities divided by mean standing stock) for biomasses in the eastern Bering Sea and in the Aleutian region are given in *Table 10*. The turnover rates in the Aleutian region are considerably higher than

85

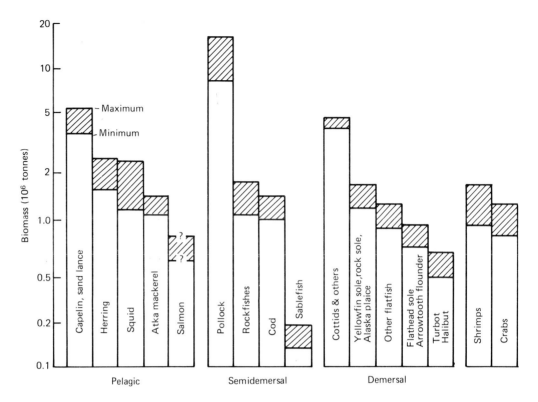

Fig 28   Equilibrium biomasses of three different regimes in the eastern Bering Sea.

in the eastern Bering Sea. The main reason for this is higher growth rates in the warmer temperatures in the Aleutian region. It is also partly due to higher utilization of food resources on the narrow shelf by seasonally migrating semidemersal and pelagic fish that migrate to feed on the shelf in summer.

The maximum equilibrium biomasses in the eastern Bering Sea and in the Aleutian region were determined also in the 'natural state' (*ie* without fishery, *Table 11*). The predation by marine birds and mammals was assumed to be the same as at present and the growth rate of exploited species was, on average, 11% lower than in the exploited state (*re* rejuvenation of populations due to the fishery, see Section 5.1 Chapter 5).

The biomass of pollock is considerably lower in the natural state. This is apparently due to the presence of a larger number of older more cannibalistic pollock in the natural state which suppresses the biomass of juveniles (see Chapter 11, *Fig 43*).

Yellowfin sole and other flatfish biomasses are lower in the natural state than in the present, fished state. This situation might seem to be contrary to common expectations, but corresponds well to happenings in the North Sea where groundfish biomasses increased considerably in the 1960s and 70s with the increase of exploitation. The larger biomasses of squids, cottids, elasmobranchs, and older populations of semidemersal fish

Table 10
Turnover ratios of biomasses in the eastern Bering Sea and in the Aleutian region. (Maximum equilibrium biomasses).

Species/ecological group designation	Turnover rate	
	Eastern Bering Sea	*Aleutian region*
Halibut	0·44	0·52
Flathead sole	0·46	0·55
Yellowfin sole	0·39	0·52
Other flatfish	0·57	0·64
Cottids	0·65	0·72
Cod	0·71	0·82
Sablefish	0·54	0·60
Pollock	0·64	0·77
Rockfish	0·56	0·47
Herring	0·70	0·70
Capelin	0·72	1·05
Mackerel	0·59	0·35
Salmon	(0·61)	(0·47)
Squid	1·60	1·60
Crab	0·34	0·32
Shrimp	0·77	0·85
Predatory benthos	0·63	0·79
Infauna	1·38	0·85
Epifauna	0·98	1·37
Zooplankton	(2·60)	(1·39)

(pollock, cod) exercise a higher predation pressure on the juveniles of flatfishes in the natural state and suppress therewith their biomasses. The production of finfish is in general somewhat lower in the natural state than in the fished state because in the former the biomasses are somewhat older with consequently lower growth rates.

Table 12 gives the maximum equilibrium biomasses in kg/km² in two shallow (continental shelf) areas (areas 1 and 4) and in two deep areas (areas 3 and 5) (see *Fig 6*). The concentration of biomasses is considerably greater in the continental shelf areas than over the deep water, with the exception of squids (also pollock in area 3). The open, narrow continental shelf of the Aleutian Chain is visited by oceanic squids. The deep water of the central Bering Sea seems also to contain high quantities of squids, which might be considered as seasonally migrating oceanic squids aggregating near their environmental (temperature) boundary of distribution. There are also differences in concentrations of individual species between different shallow areas – *eg* rockfish and Atka mackerel concentrations are considerably higher in area 4 (Aleutian Chain) than in area 1 (in the eastern Bering Sea). The small quantities of demersal fish over deep water can be considered to consist of juveniles and of a small quantity of adults in deeper parts of the continental slope.

The benthos on the continental shelf serves as an important food source for the

Table 11

Maximum equilibrium biomasses in the eastern Bering Sea and in the Aleutian region with present fishery and in the 'natural state' (without fishery). (1,000 tonnes).

Species/ecological group designation	Eastern Bering Sea		Aleutian region	
	With present fishery	No fishery	With present fishery	No fishery
Halibut	585	505	107	107
Flathead sole	875	750	127	126
Yellowfin sole	1,660	1,050	235	208
Other flatfish	1,160	1,165	187	182
Cottids	4,438	4,750	787	917
Cod	1,468	1,370	297	329
Sablefish	183	124	32	27
Pollock	15,165	11,920	6,234	5,940
Rockfish	1,825	1,660	768	790
Herring	2,327	2,215	705	775
Capelin	5,149	4,965	1,430	1,860
Mackerel	1,438	1,640	1,671	1,635
Salmon	(73)	(153)	(62)	(116)
Squid	2,310	3,030	2,695	3,610
Crab	1,225	1,105	190	165
Shrimp	1,792	1,985	241	265
Predatory benthos	818	900	99	116
Infauna	24,219	33,125	2,838	3,450
Epifauna	20,947	25,570	1,687	2,900
Zooplankton	58,430	58,430	37,230	37,230

fish biomasses in this area. Therefore, the biomasses of fish on the continental shelf areas are in general considerably higher than over deep water. On the other hand, many pelagic fish and juveniles of semidemersal fish, that depend on euphausiids as a food source, spend part of their life feeding in offshore areas where euphausiids are plentiful. It has to be borne in mind in evaluating standing stocks of fish against standing stocks of zooplankton and benthos, that the latter two are not the only food sources for fish, and that most adult fish are feeding to a large extent on other juvenile fish, including their own offspring (cannibalism).

The four-dimensional DYNUMES simulation can produce the distribution of any given species or group of species in space and time. A somewhat smoothed distribution of Pacific herring in the eastern Bering Sea during February, computed with DYNUMES simulation, is given in *Fig 29*. The model estimates of equilibrium biomass of herring in the eastern Bering Sea is 2·75 million tons (note the DYNUMES area is larger than PROBUB area, see *Fig 6*). The magnitude of annual fluctuation of this biomass is about 0·3 million tonnes. Shaboneev (1965) found the biomass of wintering herring north and northwest of the Pribilof Islands to be 2·16 million tonnes, which compares favorably with our model results. For comparison, the biomass of the herring in the North Sea has been estimated by Andersen and Ursin (1977) to be 1·8 million tonnes at the end of

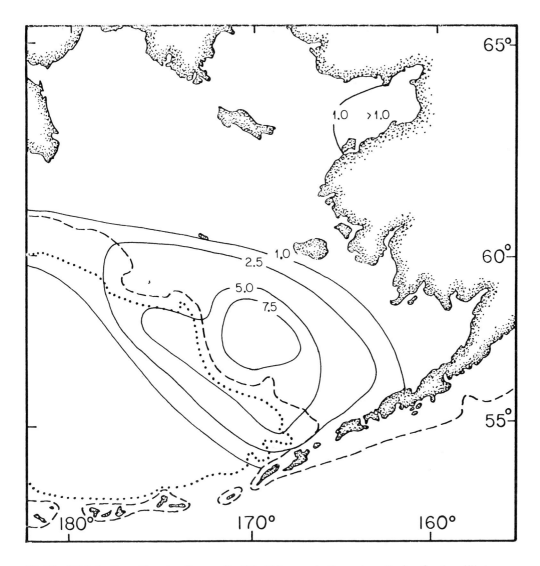

Fig 29 Distribution of herring (tonnes/km²) in February in the eastern Bering Sea (equilibrium biomass (EB), 2·75 million tonnes).

1959. The area of the North Sea is about half that of the Bering Sea included in DYNUMES; thus the biomass of Atlantic herring in the North Sea is somewhat higher per unit area than the biomass of Pacific herring in the Bering Sea.

The examples in this Chapter were meant to illustrate what can be obtained from PROBUB and DYNUMES models with respect to the evaluation of fishery resources. Detailed descriptions of the results of model applications are usually reproduced in NWAFC technical reports that have limited distribution.

Table 12
Maximum equilibrium biomasses in kg/km^2 in continental shelf areas (areas 1 and 4, see *Fig 6*) and over deep water (areas 3 and 5).

Species/ecological group designation	Eastern Bering Sea		Aleutian region	
	Area 1 (continental shelf)	Area 3 (deep water)	Area 4 (continental shelf)	Area 5 (deep water)
Halibut	724	172	726	68
Flathead sole	1,081	125	989	65
Yellowfin sole	2,088	288	1,782	125
Other flatfish	1,232	259	1,631	71
Cottids	5,720	677	6,515	333
Cod	1,628	742	1,499	258
Sablefish	204	74	213	20
Pollock	14,001	19,619	18,874	7,177
Rockfish	1,797	1,367	3,160	691
Herring	2,118	1,263	3,483	516
Capelin	4,404	3,448	5,829	1,433
Mackerel	1,325	2,326	3,060	1,599
Salmon	(20)	(20)	(65)	(70)
Squid	695	7,851	3,251	3,381
Crab	1,305	97	1,396	70
Shrimp	2,090	303	2,005	105
Predatory benthos	926	148	925	49
Infauna	28,824	900	21,657	553
Epifauna	26,296	892	21,577	365
Zooplankton	(30,000)	(40,000)	(40,000)	(30,000)

10

Environment – biota and interspecies interactions

Most of the pronounced environment-biota interactions must be included in ecosystem simulations to reproduce the ecosystem in a realistic manner. Some fixed environmental data, such as depth, are used directly as a criterion for seasonal migrations of flatfish, abundance of benthos, and in other distribution determinations. Variable environmental parameters such as temperature are used to study the effects of environmental anomalies on the fluctuations of biomasses.

The distributional changes caused by currents (as a transport mechanism) and migrations affect the distribution of fish and consequently the predator-prey relations and availability of proper food. Trophodynamics (feeding relations) and growth variations have been recognized as probably the most important aspects of interspecies interactions in the marine ecosystem (*eg* Andersen and Ursin, 1977). The interactions between growth and predation determine largely the source and sink regions of a given species. (In the source region, growth exceeds predation and other mortality; in sink regions, predation and other mortality exceed growth.) The temperature affects concurrently the growth, uptake of food, and activity, including migration.

The ecosystem simulation provides a long sought means of evaluating the environment-biota interactions and the quantitative evaluation of the effects of environmental anomalies at all space and time scales, including the study of the effects of climatic changes. The space-time changes of the distribution of environmental variables must be analyzed with special environmental models. Complete environmental models, such as hydrodynamical-numerical models, require a large computer core and considerable computer time. Thus it is difficult to run these environmental models simultaneously with ecosystem models. The environmental data fields must either be prescribed in digital form from preanalyzed data (*eg* monthly means) or environmental models must be run separately, storing their outputs on tapes or on discs in required time intervals from where they are read into ecosystem models in desired time steps.

Coupling of different simulations of biological subjects can also be performed, such as coupling of marine mammal and bird models and/or separate plankton and benthos dynamics simulation models with holistic ecosystem simulations which emphasize nekton ecosystems. The main coupling in such cases is through predation. Properly constructed dynamical, time-dependent ecosystem models can use existing single-process and/or single-species models or parts of them as adapted integral parts of holistic simulation models. Exceptions from the above role occur in those holistic models which do not have a diagnostic phase (initial analysis).

In the DYNUMES model a given biomass increases at a given location and time if its growth exceeds losses (consumption and other mortality) and decreases if losses exceed growth. These biomass increases/decreases are computed and displayed in three-dimensional (two space and one time dimension) outputs. Source (growth exceeds losses) and sink (losses exceed growth) regions have been mapped monthly (increase/decrease in tonnes/km²/month). The displays show that these biomass changes follow some relatively orderly spatial and temporal patterns. A few statements pertinent to the computation of these charts would facilitate their interpretation. The model does not use isometric growth but initial growth rates, which were determined for each species from empirical data. The growth rates are affected by age composition of the species, food availability, and water temperature. As the time step is relatively short, second-order terms in growth and mortality are neglected because other possible errors exceed these relatively small second-order terms.

The sources and sinks of all species change in space and time due to spatial and temporal changes of the processes which cause them. There is usually a sink at the periphery of the distribution of the biomass. This sink is usually compensated and maintained by outmigration from the center of the main distribution (spreading).

Two basically different causes for seasonal dynamics of biomasses, *ie* changes in abundance and distributions in space and time, can be recognized. The first group of changes are caused by seasonally changing growth, predation (and other mortalities), and production and release of eggs and milt. The second cause of seasonal dynamics is caused by seasonal migrations of species. The effects of both major causes of seasonal dynamics must be viewed spatially. Unfortunately little consideration has been given to the spatial aspects of biomass (and ecosystem) dynamics in the past, mainly because of difficulties imposed by nonsynoptic resource surveys. However, the gridded ecosystem models with spatial resolution make these studies possible. The spatial and temporal source-sink maps provide useful information on many scientific as well as practical fisheries management considerations.

March and July source and sink maps for herring in the eastern Bering Sea are presented in *Fig 30*. In winter, (March, *Fig 30A*), overall losses exceeded increases in the herring biomass. However, source (increase) areas are found during this season in the southern part of the Bering Sea near the continental slope and over the deep water seaward of the slope. In summer (July, *Fig 30B*), source areas predominate, the sink of biomass occurring only on the periphery of the main distribution. An additional example of spatial and temporal aspects of biomass dynamics is shown in *Fig 31*, which depicts the biomass sources and sinks in February and August of juvenile pollock (< 22cm long) in the eastern Bering Sea. These and other related maps show that there is a quasi-

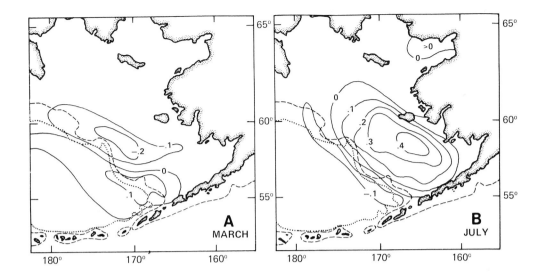

Fig 30 Sources and sinks of herring in the eastern Bering Sea, tonnes/km^2, A – March, B – July.

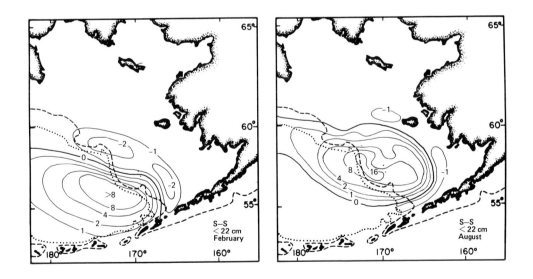

Fig 31 Sources and sinks of juvenile pollock (< 22 cm long) in February and in August in the eastern Bering Sea (in 100 kg/km^2).

93

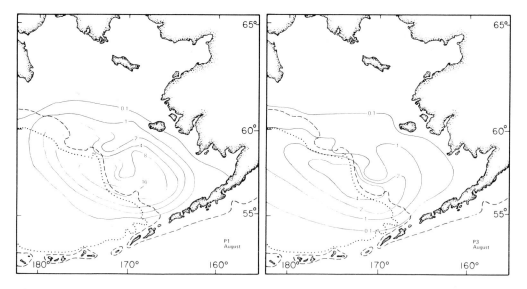

Fig 32 Distribution of juvenile pollock (< 22 cm long) and old pollock (> 45 cm long) in August in the eastern Bering Sea (tonnes/km²).

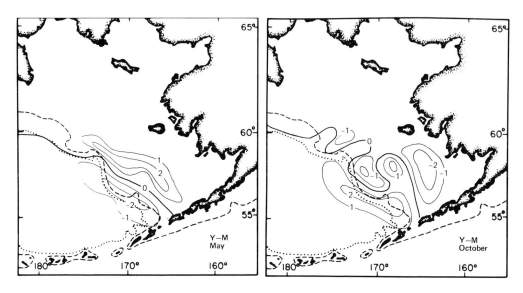

Fig 33 Changes of yellowfin sole biomass distribution due to migrations in May and in October (tonnes/km²).

continuous source of pollock off the continental slope over the deep water in the Bering Sea. During the winter this source area is further southwest where the temperature of the water is higher, allowing higher growth rates.

The distribution of two different age groups of pollock in August is shown in *Fig 32*. A partial separation of juvenile and old pollock shown with these figures is brought about, among other factors, by cannibalistic predation of old pollock on its own juveniles. The highest concentration of biomass of older pollock is found off the continental slope, whereas the juveniles are found mainly on the continental shelf.

The effects of seasonal depth migrations of yellowfin sole on its distribution are shown in *Fig 33*. Seasonal depth migrations of flatfish were investigated by Alverson (1960). On the basis of his work, it was assumed for the simulation that yellowfin sole migrates from deep water to shallow water during May and June and back to deep water in October and November. A migration speed of 3km/day was assumed in the model. The distribution of yellowfin sole in August is given in *Fig 34* as computed with the DYNUMES model.

The seasonal migrations have profound effects on other biota via trophic relations as well as on the evaluation of fishery resources with trawling surveys with respect to timing of the surveys. For example, the flatfish is dependent on benthos as a food source. The migrations cause heavy grazing of benthos in some areas during some seasons where high concentrations of fish occur. Recovery of the food resources can be expected to occur when the predators have moved out from heavily grazed areas in some seasons. The effects of spatial distribution of different prey items on the composition of food of a predator are schematically shown in *Fig 23*, which depicts a vertical section of predator-prey distributions and demonstrates that not only the food composition of the predator varies in space, but the predation pressure on the prey items varies as well.

Seasonal migrations are affected by various environmental anomalies such as temperature and current anomalies. At high latitudes, such as in the Bering Sea, water temperature anomalies are usually the most pronounced and the easiest to observe. Furthermore, more is known of the effects of temperature on fish species than of the effects of other environmental variables. Two dynamic effects of temperature anomalies are included in the DYNUMES model: the 'forced' migration of most species out of areas with subzero bottom temperatures (including a slightly increased mortality which occurs on the Bering Sea shelf), as well as the effect of temperature on food uptake and growth.

An example of the effect of water temperature anomalies on the growth of the herring biomass is given in *Fig 35* which indicates the sources and sinks of the biomass for a February with average or normal water temperature and for a February with a 1·5°C positive temperature anomaly. The growth of biomass is considerably enhanced in the latter case, especially in the southern, warmer part of the area. The effects of cold anomalies on growth are less than the effects of warm anomalies, as growth is nearly arrested at low temperatures and any lowering of temperature has no further effect on growth. However, temperature anomalies of cold bottom water have considerable effect on seasonal migrations of flatfishes to feeding and spawning grounds in shallower water. In years with extensive cold bottom water formation on the Bering Sea shelf in winter, the spring migrations of flatfishes to shallower water can be considerably delayed (Best, 1979).

In this chapter a few examples of environment-biota and interspecies interactions have been presented which can be quantitatively studied with ecosystem simulation models. Studies of many other interactions, such as the effects of currents on transport of organisms and their aggregation/dispersal, are planned or in progress. Most of the environment-biota interactions are regional and local and are studied with locally adapted simulation models.

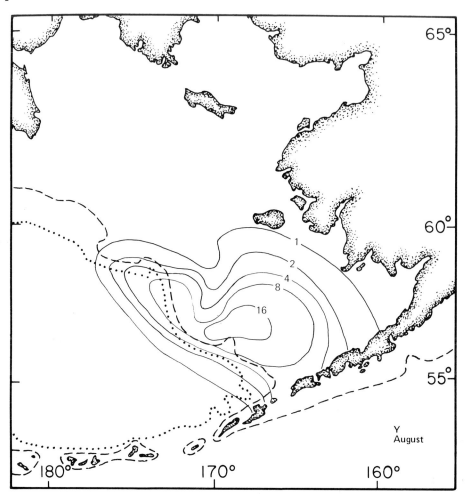

Fig 34 Distribution of yellowfin sole in August in the eastern Bering Sea (tonnes/km²).

96

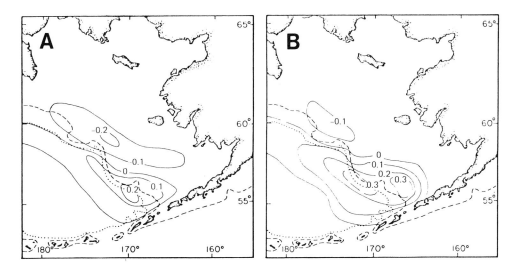

Fig 35   Effect of temperature anomaly on the source and sink of herring in February in the eastern Bering Sea (A–'normal' February, B–February with a +1·5°C temperature anomaly).

11

Natural fluctuations in the marine ecosystem

We know from empirical data and historical records that the biomasses of individual species of fish vary not only from year to year (caused *eg* by variations of year-class strengths as induced by a variety of causes) but also over long periods of time. Whereas the year-to-year fluctuations of biomasses are buffered by the presence of many year-classes, especially in longer-lived species, the long-term fluctuations can have considerable magnitudes in biomass abundance even in long-lived species. Thus, before we can properly and fully evaluate the effects of the year-to-year changes of fishery on the biomasses, we must know the causes, magnitudes, and periods of the natural fluctuations, which can and do occur without the effects of the fishery. Model studies of long-term fluctuations in fish stocks as to their causes, magnitudes and periods are few indeed. This chapter presents initial background and preliminary results of such studies with PROBUB and DYNUMES models.

In fisheries management a term 'recovery of a stock' has been used but unfortunately next to nothing is known of the recovery process or the 'rate of recovery'. Studies of natural fluctuations presented in this chapter will shed some light on this subject.

Medium- and long-term fluctuations in the marine ecosystem can occur only if the ecosystem is unstable. Thus, for the studies of these fluctuations, unstable ecosystem simulation models are required. Unstable simulation models are, however, difficult to design and require long computer programs. Furthermore, many if not most fluctuations are associated with spatial changes and their study requires models with spatial resolution, such as exist in the DYNUMES model. Both of our main ecosystem simulation models – PROBUB and DYNUMES – are unstable. The main stability-controlling processes are the so-called 'density-dependent' processes, such as predation and recruitment.

Examples of changes in distributions of biomasses, caused by seasonal migrations and by species-environment and interspecies interactions, have been given in previous chapters; it was also emphasized that the results of these seasonal dynamics must be viewed spatially. Little consideration has been given to the spatial aspects of biomass and ecosystem dynamics in the past because data from nonsynoptic resource surveys are inadequate for this purpose. However, the gridded ecosystem models having spatial resolution, such as DYNUMES, make such studies possible.

Spatially and temporally changing predation of juvenile fish is one of the major causes of long-term fluctuations of the biomasses and results in profound changes in recruitment. If some 'normal' and/or preferred prey items are not available for a given species as predator substitution with other, comparably sized food items is included in the DYNUMES model where it affects the interspecies interactions in a 'density-dependent' manner. There can also be areas and times when not enough food is available; thus starvation, with several of its consequences, will occur, the first of which is usually a delay of sex products development, affecting the larval recruitment.

The various possible types of responses of predator and prey biomasses are shown in *Fig 36*. These responses apply to both short and long time intervals. Although these

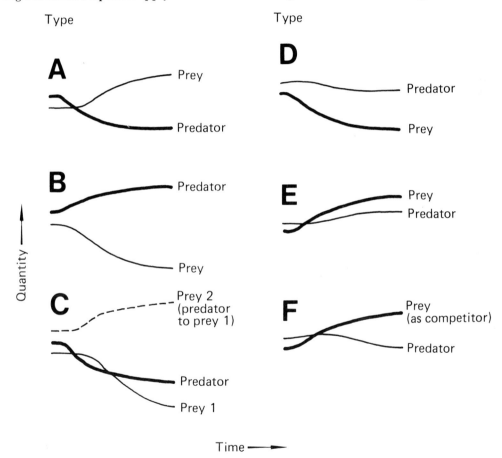

Fig 36 Types of responses to biomass changes in predator-prey controlled ecosystem.

responses can be classified into some defined types, there is considerable overlap of these responses in reality due to various predator-prey dynamic processes (*ie* changes in space and time). Some of the response types are as follows: when the predator biomass decreases, local prey abundance increases (type A), and when predator biomass increases, prey decreases (type B); when the secondary prey is predator to primary prey (*ie* secondary prey is competitor to primary predator), the predator biomass decreases (types C and F); and, if the prey is mobile and decreases by outmigration, the predator biomass might decrease as well because of forced outmigration in search of food or because of starvation (type D) and vice versa (type E).

Numerous plankton studies have shown that the abundance and availability of plankton, especially zooplankton, can vary from year-to-year (Colebrook, 1978). Thus, it can be postulated that the availability of zooplankton as a basic food item might affect the abundance of biomasses of many species. Jakobsson (1978) has postulated and demonstrated that variation in zooplankton density in the Norwegian Sea might be a cause for the drastic changes in Atlantic herring biomasses in the late 1960s.

The monthly mean zooplankton standing stock was simulated in the models, using past quantitative data on its abundance and seasonal changes that varied spatially and temporally between 400mg/m³ and 800mg/m³. Examples of the percentage of mean monthly zooplankton standing stock consumed per month in the eastern Bering Sea presented in *Fig 37* indicate that the areas of high zooplankton consumption change from month to month as affected by the distribution of consumers. These spatial and temporal changes of high consumption areas to low consumption areas in other months allow replenishments of the plankton by growth and advection in previously heavily grazed areas. Further, it can be assumed that fish eggs and larvae are consumed approximately at the same rate as zooplankton relative to their densities; thus, if an area with intense consumption of zooplankton coincides with high abundance of pelagic eggs and larvae, low larval survival and low recruitment can be expected.

Zooplankton consumption in the northern part of the Bering Sea is low, thus a great part of the deceased zooplankton can settle to the bottom, providing detritus for the abundant benthos. The consumption of benthos is also low in the northern part of the area although its standing crop is high, resulting from accumulations of generations of old epifauna. Thus, neither the benthic biomass nor the zooplankton biomass are proportional to the production of fish in all areas.

The zooplankton consumption near the shelf edge is high (see *Fig 37*). This raises several questions about the possible source of zooplankton there, as local production might be grazed down in a few months. First, the zooplankton simulation in the model for this area might be too low although considerably higher quantities were simulated than indicated in the literature. This was done because present quantitative zooplankton sampling methods are deficient (Laevastu, Dunn, and Favorite, 1976), especially in catching euphausiids. Second, the zooplankton, especially euphausiids, might be transported by currents to the convergence zone at the continental slope.

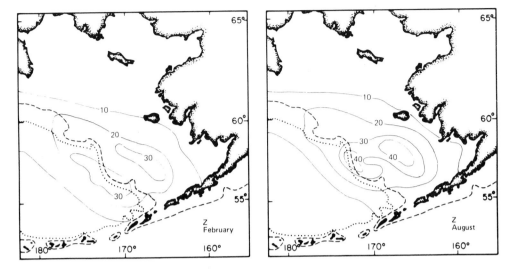

Fig 37 Percentage of monthly mean zooplankton standing stock consumed in February and in August in the eastern Bering Sea.

The studies of the long-term fluctuations of fish biomasses in a given region as caused by long-term fluctuations of basic food – zooplankton and benthos – would be realistic using the ecosystem models in areas where the long-term fluctuations of zooplankton have been studied (Colebrook, 1978). In other areas such studies can be conducted by assuming changes in zooplankton or benthos biomasses and studying the response of the ecosystem, specially its carrying capacity to such relative changes as compared to a control computational run.

The long-term dynamics of the fish biomasses in the marine ecosystem are studied with the DYNUMES and PROBUB models after determination of the equilibrium biomasses by introducing a cause for change in behavior or abundance of any species in the ecosystem. The results of such studies in terms of absolute biomass values have limited reliability beyond a few years because of the uncertainty in predicting spawning success.

Examples of monthly changes of the biomasses of shorter-lived-species, squids and capelin, are shown in *Fig 38*. *Figures 39* to *42* give examples of annual rates of changes of biomasses of some species in the eastern Bering Sea and Gulf of Alaska as computed with the PROBUB simulation model. These fluctuation studies constitute a major application of ecosystem simulation for fisheries management purposes. Since these studies are in progress, and as the results vary somewhat from one region to another, only a few general observations on the long-term fluctuations of biomasses are given below.

In past fisheries research we have been accustomed to studying the variations of abundance of individual year-classes. These variations can have considerable magnitudes

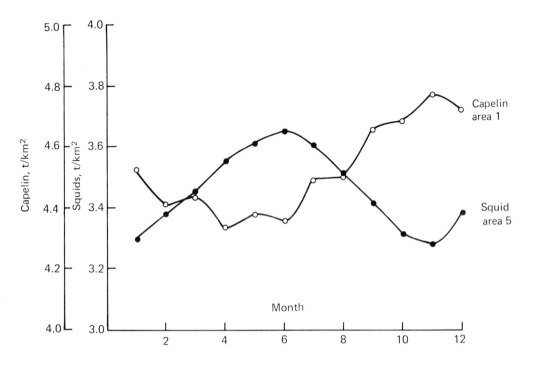

Fig 38 Examples of monthly changes of the biomasses of capelin and other pelagic fish in area 1 and squids in area 5 (tonnes/km²).

in some species. However, the biomasses are buffered with the presence of many year-classes; thus the fluctuations of total biomasses of a given species are usually considerably smaller than the fluctuations of individual year-classes. In general, the biomasses of long-lived species are more stable than those of short-lived species and the demersal species are also more stable than pelagic species.

The biomass changes presented in *Figs 39* to *42* were initiated with a sudden lowering or rising of the biomass of a major predator. Thus, the rates of changes of biomasses might represent maximum rates of changes. Recall here that the biomass change depends on the balance between growth (G) and total mortality (Z) (*ie* G-Z), and that the predation mortality is the major component of total mortality, thus, the rate of change of a biomass depends on how the predation changes with the change of the density of prey. Consequently the interspecies interactions via predation affect largely the biomass fluctuations. Also, the fluctuations are affected by year-to-year variations in environmental anomalies and differences in migration patterns and timing, affecting the predator-prey encounter.

The variations in larval recruitment will also affect the long-term fluctuations. These variations are difficult to prognosticate; only upper and lower boundaries of recruitment variations could be computed, based on past observations. In the com-

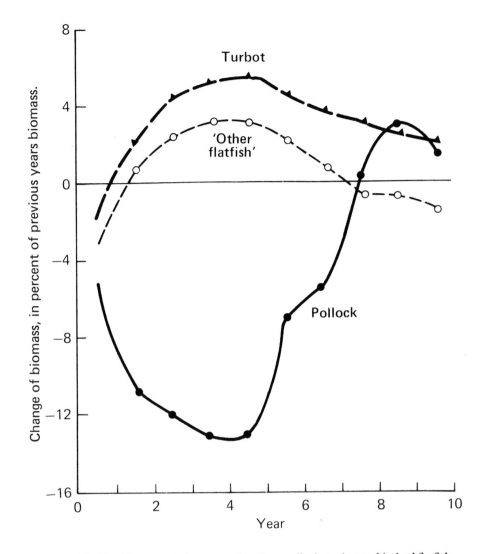

Fig 39 Changes of biomasses of walleye pollock, turbot and 'other' flatfish (several flatfish species, excluding turbot and yellowfin sole) in region 6 in western Gulf of Alaska (see *Fig 6A*) in percentage of previous year biomass. Change induced by 35% lowering of Pacific cod biomass in year 0.

putations presented in *Figs 39* to *42*, the larval recruitment was made directly proportional to the biomass of spawners.

The periods of the natural fluctuations can vary from a few years to more than a few decades and the magnitudes can be of considerable extent (*eg* the biomass can be a fraction of a few tenths to several times its long-term mean value). Many fluctuations can have irregular periods. More regular periods are caused by cannibalistic interactions

103

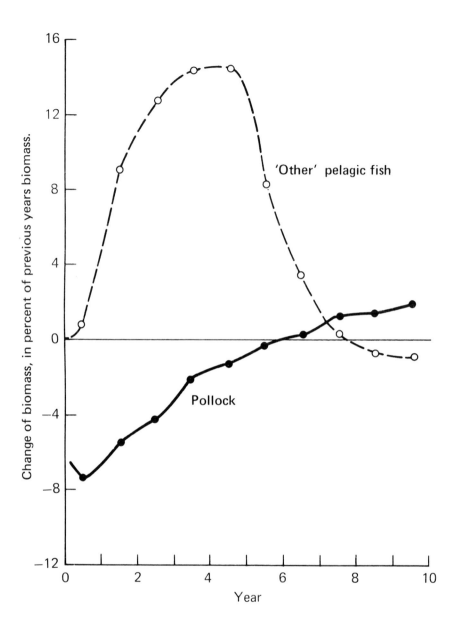

Fig 40 Changes of biomasses of walleye pollock and 'other' pelagic fish (capelin, *etc*) in region 3 in eastern Bering Sea (see *Fig 6A*) in percentage of previous year's biomass. Changes induced by 35% lowering of Pacific cod biomass in year 0.

104

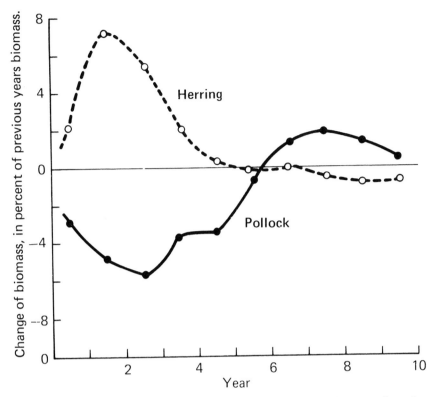

Fig 41 Changes of walleye pollock and Pacific herring biomasses in region 6 in western Gulf of Alaska (see *Fig 6A*) in percentage of previous year's biomass. Changes induced by 25% lowering of squid biomass in year 0.

in species where pronounced cannibalism occurs. The larger the fluctuation, the longer the period in species with medium to long life lengths.

The fluctuations presented in *Figs 39* to *41* have periods ranging from about 10 to greater than 20 years. Pollock, a semidemersal species, fluctuates inversely to herring and other pelagic fish. Rates of change and the periods of fluctuations presented in *Figs 39* to *41* are quite comparable to the changes in biomasses of gadoids and pelagic fish as observed in the North Sea in the 1960s and 1970s (Cushing, 1976; Ursin, 1979). Haddock in the North Sea seem to behave similarly to pollock in the North Pacific.

Long-term changes in most individual members of the ecosystem are rather complex and are dependent on changes in abundance and distribution of any one of the components of this system, biological and environmental. A complete, prognostic unstable ecosystem model, such as DYNUMES, is a prerequisite for realistic long-term fluctuation studies. There are many possible causes for the long-term changes and the need for long computer runs make these studies costly. Some of the major causes of

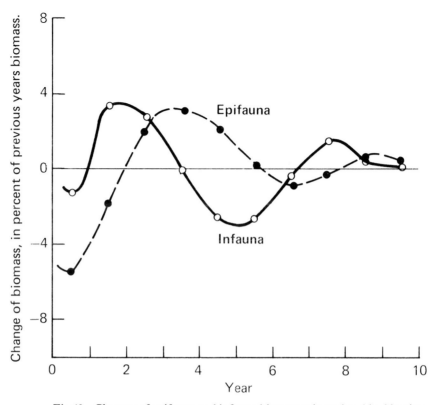

Fig 42 Changes of epifauna and infauna biomasses in region 4 in Aleutian area (see *Fig 6A*) in percentage of previous year's biomass.

long-term changes are: (*a*) increase or decrease in consumption of a given species caused by change in abundance or distribution of its predators, and changes in abundance and distribution of prey; (*b*) environmental anomalies affecting changes of growth rates and availability of food; (*c*) fishery, affecting the abundance of spawners and predation; and, (*d*) fluctuations in spawning success and larval survival.

The effect of pronounced cannibalism in older pollock on the dynamics of pollock biomasses and the effect of the fishery on pollock was studied with the DYNUMES model (Laevastu and Favorite, 1976). This study indicated that the pollock biomass has a natural fluctuation period of about 12 years. This periodicity in pollock is brought about by the cannibalism: when the older, cannibalistic pollock biomass is high, a high predation pressure is exercised on juveniles, which results in small year-classes. When the older, abundant year-classes disappear from the biomass and the earlier juveniles whose year-classes were small, become the old pollock, the predation on juveniles becomes low, because of the decreased numbers of predators present and strong pollock year-classes

106

result. The fishery affects this cycle by removing older, cannibalistic pollock and reducing the predation on juveniles (*Fig 43*).

It was also found in this study that there is an inverse relation between herring and pollock biomasses, brought about by the predation by pollock on herring. It was postulated that the increase of pollock biomass in the Bering Sea in the early 1970s might have caused the observed decline of herring biomasses.

The above described quasi-cyclic changes can occur within many fish species. Another example of this kind was pointed out earlier in respect to squids whose biomass varies from year to year partly because of recruitment fluctuations and partly because of annual migration anomalies.

All long-term changes in the marine ecosystem are time-dependent processes.

Their rates of changes and magnitudes of fluctuations depend also on the state of biomass in relation to the equilibrium biomass at the time of the forcing action (*ie* during initial influence of the fluctuation-causing event). Long-term changes can be initiated with individual alterations in most components of the ecosystem by cannibalistic interactions, successive low year-classes, and other factors. The magnitudes and period of long-term cycles are affected by the speed of recovery, which in turn is often affected by the same factors operating in opposite direction. We are not always in a position to quantify the cause of ecosystem fluctuations. Thus, the results of the determination of magnitudes and periods of these fluctuations with the simulation model gives only relative values. The factors controlling these fluctuations in unstable models are so called 'density-dependent functions', such as density-dependent

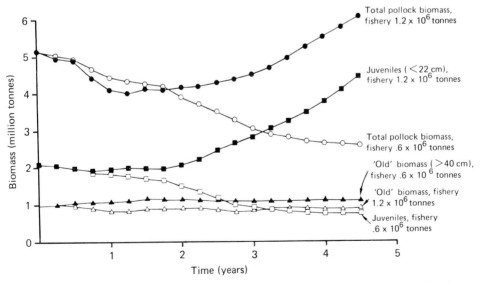

Fig 43 Change of biomass of juvenile, old, and total walleye pollock in the eastern Bering Sea with simulated present fishing intensity and half present intensity (DYNUMES submodel).

predation. It has been difficult to find empirical data for proper verification of these functions.

The simulated fluctuations give us, however, quantitative information on the rates of changes of biomasses which can, under certain conditions, be considered as 'recovery speeds'.

The rate of change of the biomass of long-lived flatfishes does not exceed 5% per year (*Fig 39*), whereas shorter-lived pelagic and semidemersal species can change up to 18% per year. The recovery rate seems to be lower than the rate of decline, perhaps because recruitment is limited. The recovery rate depends also on the state of the biomass in relation to its equilibrium level. When the biomass is low in relation to equilibrium, the recovery is slow. Recovery rates depend also on the occurrence of strong year-classes, especially several years in sequence, which cannot be predicted at present.

12

Consumption of marine biota by marine mammals in the NE Pacific Ocean

Apex predation is one of the driving functions in 'top down', predation controlled ecosystem simulation models, especially if the amounts of marine mammals are high in the region under study.

During summer the Bering Sea contains more mammals per unit area than any other ocean area. Furthermore, during the summer the numbers of marine birds in the region (over 40 million) exceed those in the rest of the northern hemisphere. Both mammals and birds are apex predators, their effect on the ecosystem is similar, but not identical to the effect of fishing by man.

The occurrence of mammals in the Bering Sea is seasonal and their distribution is uneven. In addition, their mobility can create various temporal and spatial dynamic effects on the rest of the ecosystem as a result of their grazing patterns. Estimates of numbers of marine mammals present can vary considerably from one source to another. It is, therefore, not rewarding to compute the growth of the biomass of marine mammals, as the gains in accuracy afforded by such computations would be less than that caused by errors in estimates of mammals present.

The occurrence of marine mammals and their seasonal changes in the NE Pacific are summarized in *Table 13*. There are three types of seasonal occurrence of mammals in the Bering Sea – the winter visitors (*eg* 'ice' seals and bowhead whales), summer visitors (*eg* fur seals, sea lions, shearwaters, and sperm whales), and year around residents (*eg* beluga whales).

The estimates of daily food requirements by marine mammals are also quite variable in the literature. The more conservative estimates of food requirements are between 4% to 8% body weight daily, depending on the species, and these estimates are used in the model.

The quantitative summary of consumption by marine mammals in the Bering Sea (*Table 14*) shows its magnitude and demonstrates that the marine mammal consumption can be used in the ecosystem model as the initial main driving function – *ie* to determine the equilibrium biomasses which can yield the food resource for the marine mammals, give the fishery yield, and also satisfy internal ecosystem consumption by fish biota.

Table 13

Marine mammals in the eastern Bering Sea, Gulf of Alaska and west coast of North America

Rank no.	Name	Latin name	Grp. wt.	Avg. wt.	Est. no. in N. Pacific	Maximum and minimum numbers and months in the Bering Sea and Gulf of Alaska
	Baleen whales					
1 {	Gray whale	Eschrichtius robustus	31 {	30t	11,000(E)	8,000(7)– —
	Right whale	Balaena glacialis		50	(200) ?	50(8)– —
2 {	Fin whale	Balaenoptera physalus	36 {	50	17,000	4,000(8)– —
	Minke whale	Balaenoptera acutorostrata		9	?	3,000(8)– —
3	Bowhead whale	Balaena mysticetus		35	3,000	2,500(3)– 200(8)
4 {	Blue whale	Balaenoptera musculus	45 {	75	1,700	Aleut. 500
	Sei whale	Balaenoptera borealis		30	28,000	Alaska 1,500(8)
5	Bryde's whale	Balaenoptera edeni (brydei)		30	25,000[2]	—
	'Sperm whales'[3]					
6 {	Giant bottlenose whale	Berardius bairdi	8 {	10	10,000 ?	2,000(8)– —
	Bering Sea beaked whale	Mesoplodon steinegeri		2.5	?	600(8)– —
	'Toothed whales'[3]					
7	Sperm whale	Physeter catodon		32	200,000	20,000(8)– —
8	Humpback whale[4]	Megaptera novaengliae		25	1,400	200(6)– —
9	Beluga (white) whale	Delphinapterus leucas		2.0	60,000[1]	10,000
10 {	Killer whale	Orcinus orca	8 {	10	3,000 ?	900
	Goosebeak (or Cuvier's) whale	Ziphius cavirostris		3	500 ?	200
	Porpoises and Dolphins					
11 {	Pacific white-sided dolphin	Lagenorhynchus obliquidens	120 {	60kg	10,000(E) ?	3,000(8)– —
	Dall's porpoise	Phocoenoides dalli		140	500,000	30,000(8)– 10,000(3)

No.	Common name	Scientific name			
12	Harbour porpoise	Phocoena phocoena (vomerina)	80	20,000	5,000(8)– 4,000(3)
13	Northern right whale dolphin	Lissodelphis borealis	70	40,000	—
	Risso's dolphin	Grampus griseus	100	?	—
	Common dolphin	Delphinus delphis	85 { 55	?	—
	Bottlenose dolphin	Tursiops truncatus	100	?	—
	Pilot whale	Globicephala macrorhyncha (scammoni)	200	500 ?	—
14	Sea otter	Enhydra lutris	35	120,000	100,000
	Pinnipeds, group 1				
15	Northern fur seal	Callorhinus ursinus	45	1,400,000(E)	1,100,000(7)–200,000(3)
16	Steller (northern) sea lion	Eumetopias jubatus	350	275,000	100,000(8)– 55,000(3)
17	California sea lion	Zalophus californianus	100	110,000	—
18	Northern elephant seal	Mirounga angustirostris	900	35,000	—
19	Harbour seal	Phoca vitulina (Richardi)	50	750,000 ?	270,000
	Pinnipeds, group 2 ('ice seals')				
20	Walrus	Odobenus rosmarus	800	175,000	175,000(3)– 20,000(8)
21	Bearded seal	Erignathus barbatus	200	300,000	250,000(3)– 50,000(8)
22	Ribbon seal	Phoca fasciata	70	100,000	100,000(3)– 60,000(8)
	Larga (spotted) seal	Phoca vitulina largha	56 { 55	250,000	220,000(3)– 30,000(8)
	Ringed seal	Phoca hispida	60	up to 1 mil.	200,000(3)– 20,000(8)
23	Sharks		100	?	50,000(8)– 10,000(3)

[1] Estimated Arctic population.
[2] Rare occurrence north of 35°N.
[3] The groups 'sperm whales' and 'toothed whales' signify groupings by feeding habits.
[4] Humpback whale, although a baleen whale, has been included in the 'toothed whale' group because of its rather extensive fish diet.
(E) Eastern North Pacific.

Data sources for this table see Laevastu, Livingston and Niggol, 1980.

The consumption of fish and other marine biota by marine mammals in the eastern Bering Sea and in the Aleutian region is summarized in *Tables 14* and *15*. Although plausible average estimates of abundance of marine mammals, as well as plausible average food requirements were used in the computations, the values presented in *Tables 14* and *15* might be considered as plausible maximum consumption values because simulation results indicate that the Bering Sea and Aleutian ecosystems cannot accommodate such a high consumption unless their carrying capacity is greater than that estimated in the model (see *Table 12*, Chapter 9). With initial food requirements of mammals and estimated zooplankton and benthos production it was not possible to balance the ecosystem computations unless the originally estimated mammal food requirements were lowered by 12%. The same effect would have been achieved by lowering the number of marine mammals present by the same amount. The computed high consumption might be due to overestimation of the number of mammals present. This seems unlikely when comparing the relatively conservative estimates of the numbers given in this report with estimates available in the literature. High consumption might be due to overestimation of the food requirements by mammals. This is a likely cause. Marine mammals may consume less in the ocean than aquarium research results indicate, and might grow slower in the ocean as a consequence. Partial starvation of marine mammals in the ocean might also occur.

The total consumption of finfish by marine mammals in the eastern Bering Sea is estimated by the model to be 1.77×10^6 tonnes and in the Aleutian region 0.89×10^6 tonnes, a total of 2.66×10^6 tonnes. The corresponding quantities of commercially important species taken by mammals are 1.33×10^6; 0.72×10^6; and 2.05×10^6 tonnes. The most heavily consumed species is also the most abundant species – pollock (0.73×10^6, 0.40×10^6, and 1.13×10^6 tonnes, respectively). Flatfishes are subject to the lowest relative consumption, probably because they are benthic and least available to air-breathing mammals. Only those adapted to long and deep dives consume flatfish. The estimated consumption of salmon by mammals is 35×10^3, 21×10^3, and 56×10^3 tonnes, respectively. Squids are consumed mainly off the continental shelf (total 2.98×10^6 tonnes, or about the same consumption as finfish).

The consumption of crabs in the eastern Bering Sea is 35×10^3 tonnes and from the Aleutian region only 5×10^3 tonnes. At present crabs consumed cannot be divided into commercial and noncommercial groups. The consumption of shrimps (mostly 'non-commercial') is 132×10^3 and 20×10^3 tonnes, respectively.

The consumption of benthos by mammals in the Bering Sea is 2.01×10^6 tonnes, or slightly less than the consumption of finfish. The corresponding consumption in the Aleutian region is only 0.32×10^6 tonnes. The total consumption of zooplankton by marine mammals in the eastern Bering Sea is 2.01×10^6 tonnes, or slightly more than the consumption of finfish. The corresponding consumption in the Aleutian region is 1.33×10^6 tonnes, *ie* about 40% more than the corresponding consumption of finfish in this region.

Table 14

Consumption by marine mammals in the eastern Bering Sea and in the Aleutian Region (in 1,000 tonnes).

Species/groups of species consumed	1 Baleen whales		2 'Sperm whales'		3 Toothed whales	
	Bering Sea	Aleutian Region	Bering Sea	Aleutian Region	Bering Sea	Aleutian Region
Flatfishes	—	—	2·3	1·8	21·9	4·8
Cod and sablefish	19·0	0·2	2·2	2·1	21·3	6·5
Pollock	129·6	31·6	100·8	122·4	98·2	28·3
Herring	47·3	15·4	19·8	22·5	16·5	4·0
Salmon	—	—	3·3	3·9	10·3	2·9
Atka mackerel	16·5	5·5	38·8	45·5	2·7	2·1
Rockfish	15·2	3·8	24·4	28·2	9·5	3·6
Total commercial spp.	227·6	56·5	191·6	226·4	180·5	52·3
Other fish	75·0	30·6	25·8	27·8	58·9	14·6
Total finfish	302·6	87·1	217·4	254·1	239·3	66·9
Squids	109·5	82·6	1,209·2	1,476·0	8·3	7·0
Crab and shrimp	121·7	17·4	—	—	—	—
Benthos[1]	538·7	54·1	19·7	22·0	0·4	0·3
Zooplankton[2]	1,829·8	1,135·8	142·4	176·2	2·6	2·4

	4 Dolphins, porpoises		5 Pinnipeds I		6 Pinnipeds II 'ice seals'	
	Bering Sea	Aleutian Region	Bering Sea	Aleutian Region	Bering Sea	Aleutian Region
Flatfishes	0·3	0·2	13·6	14·0	61·2	14·0
Cod and sablefish	1·1	1·7	15·3	14·7	114·9	36·0
Pollock	2·8	3·8	218·0	154·1	182·1	57·6
Herring	0·9	1·3	35·2	23·6	12·0	2·8
Salmon	0·4	0·6	15·3	12·3	5·5	1·4
Atka mackerel	0·5	0·7	32·5	22·6	—	—
Rockfish	0·5	0·8	22·3	20·4	—	—
Total commercial spp.	6·5	9·0	352·2	261·6	375·6	111·7
Other fish	1·7	1·8	107·4	65·7	169·9	35·3
Total finfish	8·2	10·8	459·6	327·3	545·5	147·0
Squids	2·2	3·7	39·7	37·6	—	—
Crab and shrimp	0·1	—	1·1	1·0	44·0	6·8
Benthos[1]	0·4	0·5	3·2	2·9	1,448·5	240·4
Zooplankton[2]	—	—	—	—	38·5	11·2

[1]Includes epifauna and infauna.
[2]Copepods and euphausiids.

Table 15
Consumption of fish and other marine biota by mammals in the eastern Bering Sea and in the Aleutian Region (in 1,000 tonnes) (See *Fig 6*).

Species/groups of species consumed	Consumption Region 1	Region 2	Region 3	Bering Sea Total
Flatfish	53·4	37·0	9·0	99·4
Cod and sablefish	74·8	57·4	41·5	173·7
Pollock	255·0	205·4	271·1	731·5
Herring	48·6	38·5	44·6	131·7
Salmon	11·3	10·3	13·2	34·8
Atka mackerel	23·0	19·8	48·3	91·1
Rockfish	19·6	15·6	36·7	71·9
Total commercial spp.	485·6	384·0	464·4	1,334·0
Other fish	190·2	152·2	96·3	438·7
Total finfish	675·8	536·2	560·7	1,772·7
Squids	122·2	95·4	1,151·2	1,368·8
Crab and shrimp	83·7	62·4	20·7	166·8
Benthos	1,145·4	778·8	86·8	2,011·0
Zooplankton	670·1	474·3	868·8	2,013·3

	Consumption Region 4	Region 5	Aleutian Region Total	Total Bering Sea Aleutian Region
Flatfish	26·1	8·7	34·8	134·2
Cod and sablefish	25·1	36·1	61·2	234·9
Pollock	130·6	267·1	397·7	1,129·2
Herring	24·3	45·2	69·5	201·1
Salmon	9·1	12·0	21·1	55·9
Atka mackerel	18·1	58·3	76·3	167·4
Rockfish	17·4	39·5	56·7	128·7
Total commercial spp.	250·7	466·7	717·4	2,051·4
Other fish	83·7	92·2	175·8	614·5
Total finfish	334·4	342·9	893·2	2,557·9
Squids	125·2	1,481·7	1,606·9	2,975·7
Crab and shrimp	12·1	13·2	25·2	192·1
Benthos	246·4	73·9	320·3	2,331·3
Zooplankton	280·9	1,044·7	1,325·6	3,338·9

The consumption of various ecological groups of fish by ecological groups of marine mammals is summarized in *Table 14*. The baleen whales consume mostly zooplankton but also take some small schooling fish. The take of fish by baleen whales is highest off the continental slope in the Bering Sea (303×10^3 tonnes). The benthos taken by baleen whales are mainly epibenthic crustaceans (*eg* amphipods). Sperm whales take, in addition to squids, schooling pelagic fish; their fish consumption is slightly higher in the Aleutian region than in the Bering Sea due to their distribution in space.

Toothed whales feed mainly on fish over the continental shelf. Therefore, consumption of fish by these whales is highest in the eastern Bering Sea area (239×10^3 tonnes including 10×10^3 tonnes of salmon). The consumption of fish by dolphins and porpoises is insignificant ($8 \cdot 2 \times 10^3$ in the eastern Bering Sea, $10 \cdot 8 \times 10^3$ tonnes in the Aleutian region) as compared to corresponding consumption by whales and pinnipeds.

Group I pinnipeds (mainly fur seal and sea lion, which feed during the summer in the Bering Sea) consume significant amounts of finfish (460×10^3 tonnes in the Bering Sea and 327×10^3 tonnes in the Aleutian region) including salmon (15×10^3 and 12×10^3, respectively). Consumption of flatfish by this group of pinnipeds is low.

Group II pinnipeds (which feed mainly during the winter in the Bering Sea), consume a considerable amount of finfish in the Bering Sea (546×10^3 tonnes) and lesser amounts in the pelagic area north of the Aleutian Chain (147×10^3 tonnes). Because of their diving and benthic feeding habits they can take flatfishes from and near the bottom (61×10^3 tonnes from the Bering Sea), their consumption of benthos being, however, considerably higher ($1,449 \times 10^3$ tonnes from the Bering Sea).

Consumption of fish by mammals is highest during the summer when their abundance in the region is highest.

The consumption of herring by pinnipeds in Group I is obviously highest over the continental shelf in the Bering Sea where both the pinnipeds and the herring occur in greater abundance. On the other hand, the consumption of herring over deep areas in the Aleutian region is higher than over the Aleutian continental shelf. The main reason is the small size of the continental shelf there.

The nature and magnitudes of the effects of mammals on the marine ecosystem, particularly on the commercial fish resources, varies from one ecological type of mammal to another and depends on several factors which are briefly described below. The 'offshore' baleen whales feed on the same 'trophic level' as most pelagic fish and consume the offshore zooplankton, competing possibly with salmon, herring, and pollock in this regime; gray, white, and bowhead whales compete directly with many of the commercial fish on the continental shelf where they consume epibenthos. At this time it is difficult to evaluate the competition between whales and pelagic fish, as empirical data for zooplankton production, especially of euphausiids, are deficient in the North Pacific Ocean. The standing stocks and production of epibenthos in the Bering Sea is virtually unknown, mainly due to difficulties in obtaining quantitative samples. Monthly zooplankton standing stock is simulated in the model, based on best available

115

data. Less than 25% of this standing stock is consumed per month in offshore areas.

Among pinnipeds, only walrus and bearded seals consume considerable amounts of benthos. As these animals use larger benthic organisms which are unsuitable as fish food because of their size, the competition for the benthic food resource between demersal fish and these mammals is minimal. On the other hand, toothed whales and most pinnipeds feed directly on fish and, therefore, compete directly with man. Because mammals often prey on juvenile fish, this predation can affect both the size distribution of the fish and the size of the exploitable biomass of the fish (*Fig 44*).

Conversely, the effects of fishing on the food of marine mammals cannot be empirically demonstrated at present. There is no empirical proof that the fishery in the Bering Sea (maximum 2·2 tonnes per km² as compared to 6·7 tonnes per km² from the North Sea) has adversely affected the total finfish biomass; in fact, there is good evidence that pollock, cod, herring, and yellowfin sole biomasses have increased substantially during the last decade. The rejuvenation of biomasses which was a consequence of fishing, might be working in favor of marine mammals who feed predominantly on smaller fish than taken by the fishery. Although there have apparently been rather pronounced fluctuations of the biomasses of some species, there is no proof that the total finfish biomass has decreased. There is theoretical evidence, as well as empirical observations, from several areas (*eg* North Sea) that fishing does not affect the total finfish biomass, because when one species declines in abundance other species might increase with some consequent changes in trophic relationships. Exceptions to this rule are the pelagic fish ecosystems in upwelling areas such as anchoveta off Peru.

Studies of the possible effects on marine mammals by the fishery require new hypotheses as to causes and effects in addition to accurate observations on the dynamics of mammal populations. Studies of the magnitudes and periods of natural fluctuations of finfish are conducted with ecosystem simulation models to enable us to separate any effects of the fishery and mammal consumption from other natural fluctuations.

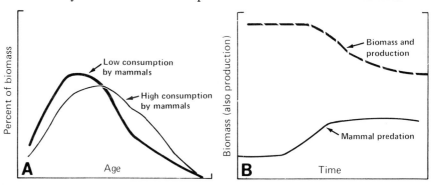

Fig 44 Schematic presentation of the effects of predation by marine mammals. A – The effect of mammal predation on the distribution of biomass of a species with age. B – The effect of mammal predation on the biomass level of total finfish biomass in an ecosystem.

116

13

Verification and validation of ecosystem simulation

Traditionally the real test of fisheries models has been how well a model 'predicts' rather than whether or not biologically sound terms have been used to describe its coefficients and exponents (Radovich, 1976). If a model did not work, something was considered wrong either with the mathematics, the assumptions, or the data. However, in ecosystem simulation tests different approaches prevail and it is necessary to differentiate between verification and validation in large ecosystem simulations.

Verification refers to checking of logic and the correctness of individual models and formulas used in the simulation which are verified with available empirical data. This part of the verification is carried out primarily during the design of simulation. Verification includes also the testing of the simulation at large, using various impulses as input, whereby the expected response of the ecosystem to the impulse must be at least qualitatively known for control purposes. The verification of the ecosystem simulation model in testing the formulas used in the models for the reproduction of known effects and behaviours for which they are designed is in many ways a continuing process. For example, the effect of water temperature on growth in the DYNUMES model was initially formulated on the basis of some early knowledge available on the subject, notably Krogh's metabolic curve. When an excellent paper by Jones and Hislop (1978) appeared later, dealing partly with the subject, verification and additional tuning were provided. Further empirical evidence on the effect of temperature on herring abundance was provided by Grainger (1978). Thus the verification of large ecosystem models also becomes a continuing study of the response of the ecosystem to changes in various rate and state parameters.

Variations in results of deterministic simulations depend also on the accuracy and reliability of the input data because the ecosystem models rely heavily on various input parameters for determination of the initial state.

The basic input data for the PROBUB 80-1 run are given in technical reports (*eg* Laevastu, Livingston and Niggol, 1980) where some of the limitations of these data are briefly discussed. Detailed discussions of the accuracy of the input data would require voluminous works as the accuracy of individual data varies from species to species and from area to area. There are some basic inputs (such as growth rates and food requirements) which affect the determination of equilibrium biomasses more than others. There are also multitudes of minor error sources which propagate through the computations, as schematically shown in *Fig 45*.

The limits of plausible errors caused by possible errors in growth coefficients can be determined with empirical data and with the biomass balance formula.

Errors in spawning stress mortality (senescent mortality) affect the equilibrium biomass relatively little, as this mortality coefficient is small and relatively constant. Errors in food requirement coefficient are limited by food availability-dependent feeding (leading to starvation) and by substitution of part of the lacking food by the 'buffer food sources' in the ecosystem (*ie* zooplankton and benthos). Overconsumption of a species is not possible in PROBUB as the limits of consumption which are derived from the data on turnover rates, fecundity, larval growth, and age-variable total mortality determination are possible. The limitations in food item availability cause changes in food composition (in computations) and lead to an overall maximum utilization of all food resources in the ecosystem. Food composition changes can be best treated in the DYNUMES model with spatial resolution which allows consideration of the predator-prey distribution overlap.

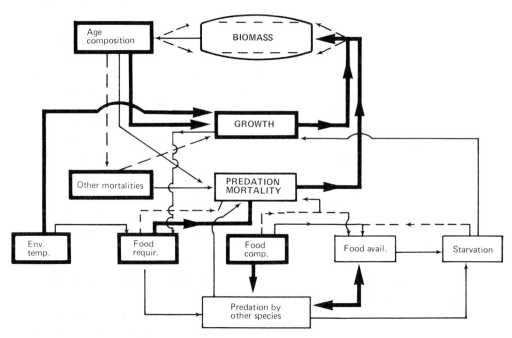

Fig 45 Schematic presentation of causal relations via processes and their feedbacks which affect the error limits of the determination of equilibrium biomasses. (Heavy lined boxes – minimum and maximum values obtainable from empirical data. Heavy influence lines – major effects; dashed lines – minor effects).

There is always some migration between the different computation areas in the PROBUB model; for example, the offshore regions are often source regions for continental shelf areas during certain time periods. Some migration is prescribed conservatively in PROBUB, and there is no reasonable way to determine the plausible error in this migration estimation.

The possible errors in recruitment simulation do not enter into the determination of equilibrium biomasses (*re* iterative adjustment of biomasses at the end of each computation year). After the equilibrium biomasses are computed, the recruitment variations are dampened by making them proportional to the square root of the relation between equilibrium biomass and actual biomass. This approach does not, however, simulate the possible effect of exceptional larval survival.

The error limits presented in *Table 16* have been derived via a multitude of ways: by determining the error limits in individual formulas, assuming plausible minimum and maximum values to input data; by following the error through the computation procedure (see *Fig 45*) and estimating the magnifications and reductions of the errors; and by conducting many numerical experiments with the model where input data were assigned different plausible values (a kind of 'sensitivity analysis'). Some remarks are given in this table as to the largest plausible sources of errors in individual ecological groups; thus these remarks also indicate further research needs and priorities.

The values in *Table 16* should be considered as upper limits, as the overall error might be smaller by indirect compensating mechanisms (such as one parameter might be lower and another higher than the plausible value). Values in *Table 16* should not be confused with an intentional combination of major data limits, such as are used in determination of maximum and minimum equilibrium biomasses (*Table 9*, Chapter 9).

An important part of the verification of traditional models is the sensitivity analysis. Sensitivity analysis indicates where the influence of possible flaws in available knowledge lead to major consequences; thus sensitivity analysis can also serve as guidance for further research. In the past, sensitivity analysis in simple few-parameter models was done by changing one constant (*vice* parameter) at a time. This method is not applicable to large ecosystem models as the number of simulations required is prohibitively large. Behrens (1978) has outlined an analytical method for sensitivity analysis of models consisting of ordinary first order differential equations, but his method is not fully applicable to ecosystem simulations either as the PROBUB and DYNUMES models are deterministic and use a variety of empirical and differential equations, the latter usually solved in finite difference form. Thus other approaches must be devised to study the sensitivity and accuracy of the model and its results. The first method is the verification of individual formulations (submodels) used in the simulation as described in previous paragraphs. This task is usually accomplished during the design of the model and during the preparation of the inputs.

The second sensitivity analysis of the simulation is conducted via a number of changes in input parameters in the process of updating the inputs. A first change might

119

Table 16

Estimated plausible maximum error limits of equilibrium biomasses in PROBUB 80-1 (preliminary, in % of plausible mean value).

Ecological group	Maximum error limits (%)	Remarks on largest plausible source of errors
Flatfishes	18	Seasonal changes in food uptake and composition, due to seasonal depth migrations
Pollock	20	Spatial change of growth rate and offshore distribution of biomass during some seasons
Herring	25	Seasonal and spatial changes in contribution of herring and rockfishes to food of other species; seasonal migrations
Rockfishes	30	
Cod, sablefish	20	Growth rates of juveniles, seasonal migrations
Other noncommercial demersal	25	Growth rates; size-age distribution; occurrence in diet of other species; age of maturity and senescent mortalities
Other noncommercial pelagic	30	
Crabs, shrimps	25	Growth rates, distribution (spec. of juveniles)

involve the updating of marine mammals and bird numbers and their food composition. The first input usually includes conservative estimates of marine mammals whereas the revision will include plausible amounts.

The third sensitivity test involves the updating of food composition of fish and the fourth sensitivity test concerns the change of the fisheries. These latter are the main tests with the model for management purposes.

In deterministic models, sensitivity analyses become studies of specific responses of the systems to expected changes of parameters. The variations in results of deterministic models depend also on the accuracy and reliability of the basic input data, the accuracy of which can be evaluated by empirical statistical methods.

Validation of simulation refers to the comparison of principal results from simulations with direct observations in the field. Usually these results present either abundance and/or distribution changes of a given species if and when a causative factor for these changes has been introduced in the simulation model. Special research projects usually provide validation of the various rate parameters.

Although the dynamic aspects of the Bering Sea ecosystem are difficult to validate empirically due to the scarcity of data, particularly from extensive time-series surveys and migration studies, the validation of general abundance and distribution results is possible by comparing the computed abundance and distribution of biomasses with independently obtained empirical data such as those obtained from fisheries surveys. This comparison is rewarding only if the survey data are converted to total exploitable biomass, using catchability coefficients. Examples of this validation are given in Table 17 for the species which are more reliably reported quantitatively in the surveys with the otter trawl. In general, resource survey estimates are considered only to be about $\pm 50\%$ accurate (Grosslein, 1976), whereas the error in the model computation results does not exceed $\pm 30\%$ of the computed value.

Qualitative validation of the simulation models can be provided by occasional, special fisheries surveys. The following serves as an example. In the early stage of the Bering Sea ecosystem modeling, it became obvious that there must be a considerable amount of pollock (and some other fish species) over the deep water in the Bering Sea. However, pollock were never caught commercially over deep water and the model results were severely criticized until a Japanese survey showed considerable amounts of older pollock over deep water. Furthermore, the model studies showed that the deep water areas are source areas of biomass for many pelagic and semipelagic species at least part of the year, the abundant euphausiids in this area providing an ample food source. However, no extensive schooling occurs over deep water, making the fishery less profitable there than over the continental shelf.

Table 17
Comparison of exploitable biomasses of some species in the eastern Bering Sea as obtained by surveys and as computed with PROBUB model (in 1,000 tonnes).

Species/Groups of Species	Mean 1975, 1976 surveys (converted from Bakkala and Smith, 1978)	Equilibrium exploitable biomass from PROBUB model	Catch 1975
Greenland turbot, halibut	176	222	65
Flathead sole, arrowtooth flounder	206	377	26
Yellowfin and rock sole, Alaska plaice	2,716[1]	509[2]	74
Pollock	3,698[3]	6,449	1,285
Cod	233[3]	746	57

[1]Possible overestimate during 1976 survey; 1976 biomass was more than twice the 1975 biomass.
[2]Possible underestimate as 1960-1970 input data were used, when yellowfin stock was low.
[3]Pollock and cod are not fully evaluated by bottom trawling on the shelf.

The examples of verification of ecosystem simulation, given in this chapter, demonstrate that the subjective verification via utilizing total accumulated knowledge of this system is of utmost importance in ecosystem simulation. It is imperative that colleagues of fisheries science and practice participate in the verification process of the models – after all it is the knowledge provided by them which is compiled in a systematic manner in the simulation.

Validation of the simulation results relies heavily on local data and surveys and is thus easier in areas of abundant fishery data. In other areas special surveys and tests must be conducted. In the validation of computed equilibrium biomasses the effects of long-term fluctuations and the present state of the ecosystem with respect to these fluctuations must be considered.

There are, however, two specific subjects in which our knowledge on nearly all species and areas is deficient; the abundance and migrations of prefishery juveniles and the biomasses of noncommercial dispersed species.

14

Simulation of the fishery in marine ecosystem models

The simplest formulation of the theory of fishing (Russell, 1931 – from Cushing, 1968) is:

$$P_2 = P_1 + G + R_r + Z \tag{55}$$

Where: P_2 is stock at time t_2; P_1 is stock at time t_1; G is growth between t_1 and t_2; Z is mortality between t_1 and t_2; and, R_r is recruitment.

Russell's theory of fishing treats the effect of fishing within the total mortality Z (mortality between t_1 and t_2). Most fisheries dynamics approaches treat the fishery as fishing intensity, following Baranov's (1918, 1926) original theory. In the single-species fisheries dynamics models it is assumed that the unit gear takes a constant proportion of the fishable population. Furthermore, it partially follows from the above assumption that fish are distributed randomly over the fishing area, that unit effort is independent of the distribution of fish and the fishery, and that a linear or quasi-linear relation exists between catch per effort and population size.

The above assumptions are seldom approximately valid and consequently a given unit of fishing effort can take an increasingly larger proportion of the fish population as the population declines. The fishing effort and catch per unit effort are not related to stock size, especially if modern fish finding methods such as sonar are used, as approximately the same effort is needed to catch the first tonne of pelagic schooling fish as to catch the last tonne of fish from the last school. Only in partially schooling, demersal fish the catch per unit effort can be roughly proportional to the population present. And then only if availability of fish in an area and migrations are taken into consideration.

In the fish ecosystem simulation models described in the foregoing chapters, no direct use has been made of fishing effort or catch per unit effort for determination of fish stocks except in some validation attempts related to survey results of demersal fish. However, the known catches are simulated in the model with proper time and space resolution. In the use of models for the study of the effects of the fishery on the ecosystem, we must, however, simulate various proposed fishing efforts.

Most single-species fisheries dynamics models have been number-based and have used one year time step. Thus, the fact that the effect of the fishery on biomass changes is not a linear function of catch, has been largely overlooked. If we catch 10 000 tonnes of a given species, the annual decline in biomass of that species due to the fishery is usually complex and not in direct proportion to the catch, as will be briefly demonstrated in this chapter. This fact has been difficult to appreciate, as it is not apparent from conventional number-based computations using an annual time step.

Biomass-based models require the computation of growth, mortality, and the fishery which are all age and biomass distribution dependent. Therefore, the effect of the fishery on the biomass does not necessarily become linearly dependent on time nor on the initial exploitable biomass. This is demonstrated with the following special model which is a simple truncated single-species, biomass-based model where predation is prescribed or estimated in a separate model.

The biomass dynamics in this simple model is computed in monthly time steps with the following well known formula:

$$B_t = B_{t-1} e^{g-z} \tag{56}$$

where t is time step (one month) and B is biomass (initial biomasses 3 600, 4 200, 4 800, and 5 400kg/km² assumed in the examples) and g is instantaneous monthly growth coefficient ($=0\cdot10$ in the example).

In this simple model the biomass can be assumed to be the biomass of a whole population, biomass of a cohort, or groups of cohorts. The coefficient Z must be adjusted correspondingly:

$$Z = F + m + M_p \tag{57}$$

where m is instantaneous mortality coefficient from old age and diseases ($=0\cdot003$ in the example), F is instantaneous mortality coefficient ($=0, 0\cdot015$, and $0\cdot0225$), and M_p is instantaneous predation mortality coefficient (prescribed in the example below either as a constant or constant plus density-dependent fishing mortality).

The instantaneous predation mortality coefficient is computed from known or assumed predation (C):

$$M_p = -\ln\left(1 - \frac{C}{B}\right) \tag{58}$$

In our example C has been assumed to be 360kg/km² per month.

Variation in predation on a given species is dependent on the density of the prey. The density-dependent function is assumed to be approximately linear. Furthermore, the predation is expected to contain also a density-independent part due to selective feeding. Thus, the following predation functions were computed with all assumed initial biomass levels:

124

$$C = \text{Constant} \tag{59}$$

and

$$C_t = A + D\frac{B_{t-1}}{B_b} \tag{60}$$

where constants A and D in the examples below are: A=100 and D=260, the constant A being density-independent predation. Density-dependent predation is represented by; actual biomass of prey (B_{t-1}) divided by the mean biomass (equilibrium biomass) (B_b), which is the initial (input) biomass.

The fully density-dependent predation was also computed as:

$$C_t = C\frac{B_{t-1}}{B_b} \tag{61}$$

The corresponding partially and fully density-dependent predation (at computation range only slightly nonlinear) functions tested were:

$$C_t = A + D\sqrt{\frac{B_{t-1}}{B_b}} \tag{62}$$

and

$$C_t = C\sqrt{\frac{B_{t-1}}{B_b}} \tag{63}$$

The coefficient m of old age and disease mortality, including spawning stress mortality, has been assumed to remain constant in the example unless the fishery is considered to operate on older year-classes and biomass recruitment to exploitable stock is smaller than yield, in which case the m should decrease and growth coefficient for the whole biomass should increase.

If the fishing mortality coefficient F remains constant, it is a fishing intensity coefficient and the catches decrease with decreasing biomass. This might be the case with trawl fishery for *eg* flatfishes.

In most cases the annual catch is known or there might be a need to simulate the effect of a given annual catch on the biomass of target species. In this case the fishing mortality coefficient is computed (tuned) on equilibrium biomass and must change if the biomass changes to reproduce the known annual catch. This also represents an example of targeted fishery on schooling species:

$$F_t = F\frac{B_b}{B_{t-1}} \tag{64}$$

Figures 46 to *48* illustrate the change of biomass with fishery, computed with a combination of density-dependent and density-independent predation and fishing mortality coefficients and other formulas described above. The computations have been made with the assumption that the recruitment is strictly proportional to biomass present.

The effect of fishery on recruitment occurs mainly when recruitment overfishing can be expected. This would be the case in an intensive fishery on small spawning biomasses and also intensive fishery on long-lived, low-fecundity species.

The relation between quantity of biomass removed by fishery and the change of biomass, is not strictly proportional to fishery (*Figs 46* to *48*). Reasons for this unproportionality in biomass decline are that mathematically growth, fishery, and predation are exponential coefficients, thus biomass change is a nonlinear function of the change of these coefficients. Furthermore, computationally the process of biomass change with time is a finite difference approach, thus dependent on the length of time step (and the biomass in previous time step). The effect (proportionality) of fishery on the changes of biomass depends also on the state of the biomass in relation to its equilibrium state (*Fig 48*). The fishery will accelerate the decline of small biomass, whereas it will slow the increase of large biomass.

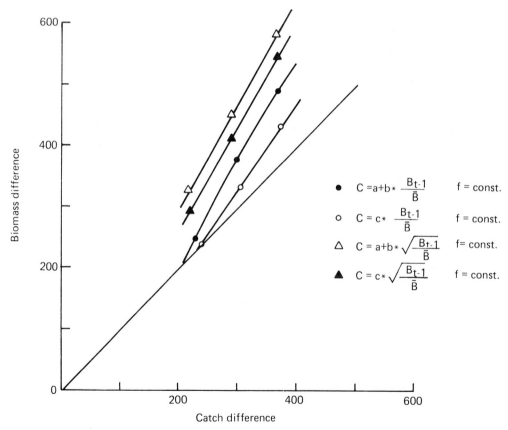

Fig 46 Difference in biomass resulting from differences in catch with various variations in consumption (predation) (constant fishing mortality coefficient).

126

The fishery does not change the density of prey (prey being the target species), as it operates on older, larger fish. However, the fishery affects predation on other species as it removes larger, more fish-eating specimens. Also, a seasonal fishery has different effects on biomass growth and decline than a fishery which is evenly distributed in time (and space), especially in higher latitudes where pronounced seasonal differences in growth occur.

Figure 48 demonstrates that the decrease or increase of the biomass with time depends on whether the growth of biomass per unit time is smaller or larger than its removal by mortality and fishery. It also demonstrates that there can be a singular amount of biomass where these two processes are quantitatively in balance (see further Chapter 9 on equilibrium biomasses). This exact balance rarely occurs in real ecosystems. Thus, these figures also indicate that marine ecosystems are basically unstable.

The main stabilizing mechanisms in the ecosystem are the density-dependent removal and addition processes (predation and recruitment). The exact quantitative form

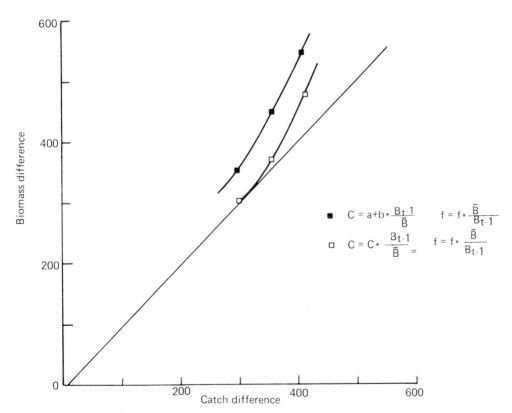

$$\blacksquare \quad C = a + b * \frac{B_{t-1}}{\bar{B}} \qquad f = f * \frac{\bar{B}}{B_{t-1}}$$

$$\square \quad C = C * \frac{B_{t-1}}{\bar{B}} = \qquad f = f * \frac{\bar{B}}{B_{t-1}}$$

Fig 47 Differences in biomass resulting from differences in catch with different consumption (predation) (constant yield).

of these density-dependent functions is not known at this time. These functions are expected to be influenced by many environmental and species specific factors and their study is one of the important research tasks in fisheries. The truncated single-species, biomass-based submodel, described in this chapter, can also play an important part in these studies.

Changes in predation are often more important in biomass change than changes in the fishery, and there is a relationship between the change of fishing mortality and the change of biomass growth coefficient (*re* rejuvenation of populations) and senescent (and spawning stress) mortality coefficient.

The effects of the fishery in the DYNUMES model are simulated by changing the fishing effort (intensity) coefficient either as to location, time, or magnitude or a combination of these changes.

The computation of yield (Y) in the ecosystem model is usually performed with fishing intensity (fishing mortality) coefficient. This coefficient F is computed for the desired time step (*eg* month, using the equilibrium biomass (B_e) and the acceptable catch estimate (ACE – see next chapter) or actual landings for a given year.

$$F = -\ln\left(1 - \frac{ACE}{12B_e}\right) \tag{65}$$

$$Y_t = B_t - B_t\,e^{-F} \tag{66}$$

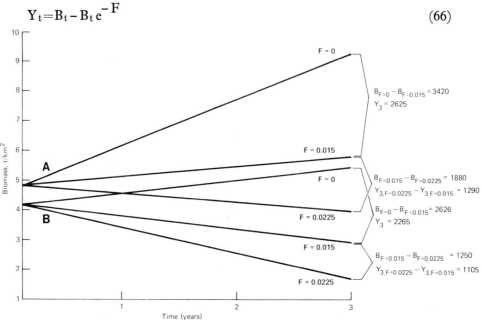

Fig 48 Changes of biomasses over 3 years with different fishing efforts and with two different initial biomasses. The recruitment is directly proportional to biomasses present and predation mortality (C) is density dependent.

In the latter formula, the yield fluctuates with the fluctuating biomass. However, if a given (constant) yield is desired, the following formula must be used:

$$Y_t = B_t - B_t\, e^{\dfrac{-FB_t}{B_e}} \tag{67}$$

Mortalities change with age; consequently, if the biomass distributions change with age due to fishery, there must be corresponding changes in mortalities. A biomass distribution is assumed to be in balance when: present fishery minus recruitment to fishery equals removal by fishery plus spawning stress and senescent mortalities.

Spawning stress mortality (M_s) operates on the exploitable part of biomass. Thus, if the fishing intensity changes, the spawning stress mortality must change. This can be approximated with the following formula:

$$M_{s,new} = M_{s,orig} + 0.25\, \frac{F_o - F_n}{F_o}\, M_{s,orig}. \tag{68}$$

where: F_o is original fishery, F_n is new fishery, and M_s is spawning stress mortality. Growth rate changes also with fishery, which can be approximated with the following formula (see Section 5.1 Chapter 5).

$$g = g_o + 0.25\, \frac{F_n - F_o}{F_o}\, f_a\, g_o \tag{69}$$

where: f_a is the fraction of exploitable biomass.

The Formulas 68 and 69 are relatively rough approximations. Theoretically exact computations would be possible for both formulas; however, such computations would require additional parameter fields to be stored in the computer and additional computation time. Both are already at a premium in large ecosystem simulations and the small additional accuracy gain is not warranted.

15

Acceptable catch and its estimation

One of the main objectives of ecosystem simulation modeling for fisheries is to determine quantitatively the response (changes) of the ecosystem to exploitation and to determine how much of any given species can be taken from the ecosystem of a given region without causing 'changes of undesirable nature and extent'. The decision of what are 'undesirable changes' in the ecosystem must be based on many criteria, most of which have little to do with science proper, but include economic considerations, such as marketability and prices of fish, *etc.* However, science must show the nature and extent of the changes and make suggestions as to which changes are acceptable from the ecosystem point of view.

The effect of the fishery on the ecosystem can be determined only in a full (complete) ecosystem simulation model where all species and all fisheries are included. Any removal of some quantities of fish will cause changes in the biomass of the target species as well as in other species not subject to the fishery. The criteria of which changes in the ecosystem are acceptable from the ecosystem 'health' point of view, considering especially its capacity to reproduce harvestable species biomasses, must usually be established on regional bases, with the knowledge that these criteria should vary in time with changing economic conditions. The establishment of these criteria are outside the scope of the ecosystem simulation task.

The estimation of the acceptable catch is not part of ecosystem simulation proper but, rather, an auxiliary approach to facilitate the application and use of the simulation model. Although in this chapter we attempt to produce with simple indices some numbers as guidance for 'acceptable catch level', we do not intend to give a definite and simple answer to a very complex problem. The first guess estimate of 'acceptable catch level', derived from the approach described below, will merely serve as first guess input of fishing intensity (and/or yield) into the ecosystem simulation to determine the response of the ecosystem to these yields and to provide background for management decisions. The first guess estimate is necessary to reduce the trial-and-error inputs for testing of the effects of fisheries with these models. The complex and complete simulation models enable evaluation of the changes in the stocks caused by natural fluctuations and fishing.

In the past, the fisheries management approaches attempted to estimate the Maximum Sustainable Yield (MSY). This term was derived from the report of the Select Committee of the House of Commons in 1893 which described the need for and the purpose of fisheries conservation as: '. . . to enable fishermen to obtain from a stock of fish the highest yield, consistent with that yield being maintained in the future . . .'. In recent years a number of closely related terms have been coined, such as optimum yield (OY), equilibrium yield (EY), allowable biological catch (ABC), and others.

To distinguish our estimated catch (quantity of possible catch from a given region) from earlier used terms, we are calling our estimated number 'Acceptable Catch Estimate' (ACE). What is an acceptable catch depends on the extent and nature of the changes in the ecosystem as a result of any catch and a great number of considerations of economic and social nature; it will usually be debated and determined by concerned groups (fishermen) and the fishing industry.

Larkin (1977) wrote an epitaph to MSY and related concepts. He has described all the objections and difficulties involved in quantification of any of these terms as well as the fallacies and wrong expectations connected with these concepts, which are neither fully biological nor economic concepts.

Managers and politicians, however, require a number or quantity which might show how much of any species they will allow to be taken by fishery from any given area. Further, mathematicians seem to love, at times, to manipulate numbers which have no meaning in the real world. These two phenomena have perpetuated the religion of MSY.

In the past a simplified approach for estimation of MSY has been used:

$$MSY = K \, M_n \, B_o \tag{70}$$

where K is assumed to be an empirical constant (mostly used value is 0·5 but can range from 0·3 to 0·6); M_n is natural mortality, which is usually an unknown quantity that can vary from 0·2 to 1·0; B_o is the size of 'virgin biomass', which is, in most cases, also an unknown quantity that can at best be estimated to the order of magnitude with available methods. The lack of information required by the formula indicates that MSY is an unknown quantity. Consequently, another method of estimating acceptable catch, which would be based on known biological and ecological concepts, must be found.

Before establishing the criteria for the estimation of ACE it might be helpful to review briefly the dynamics of the marine ecosystem in response to the fishery with the help of a few graphic illustrations. The total finfish biomass in a given region fluctuates but little in the course of time (*Fig 49*). (An exception may occur in upwelling regions.) However, biomasses of individual species can fluctuate considerably – one decreasing, the other increasing. Individual species fluctuations are not always caused by the fishery, but may be caused by other factors such as environmental anomalies. Thus, the determination of the causes, magnitudes, and periods of these natural fluctuations is one of the important tasks of modern fisheries science (see Chapter 11). The total finfish biomass (the carrying capacity of finfish) is determined by the production of organic matter, its

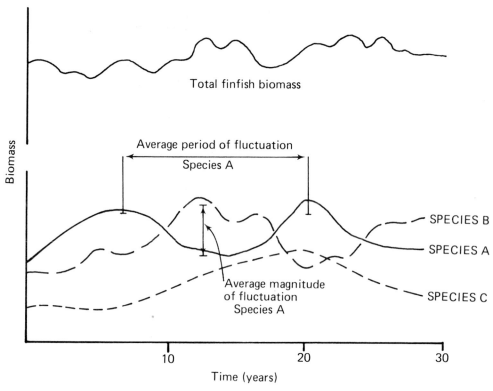

Fig 49 Schematic presentation of fluctuations of total finfish biomass and biomasses of individual species with time.

turnover, and the benthos and zooplankton. The true carrying capacity of an ecosystem is always smaller than the theoretical carrying capacity, computed with the assumption of full utilization of organic production plus the circulation of the biomass of the smaller members of the finfish community.

The following figures will help one to realize that the dynamics of a single species biomass varies considerably from species to species and that no simple concept (MSY or others) can be fully valid for all species. *Fig 50* shows the change of mortality with age (relative to the biomass present in each year-class). The high predation mortality in larvae and in young fish decreases rapidly with the growth of the fish and reaches a minimum at a given age (size) which usually coincides with the size where the fish come under the fishery. Thereafter the spawning stress mortality starts to increase with age.

Fig 51 illustrates schematically what happens to biomass distribution with age when a virgin stock comes under the fishery (assuming constant recruitment). Within about five years the older part of the biomass (exploitable biomass) which was originally

132

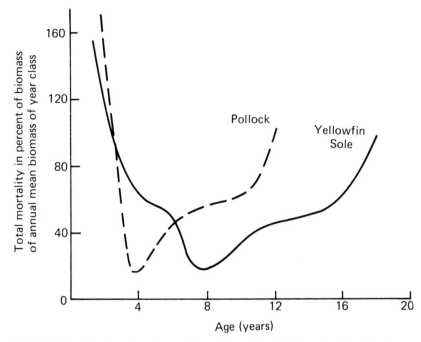

Fig 50 Distribution of total mortality with age in pollock and yellowfin sole.

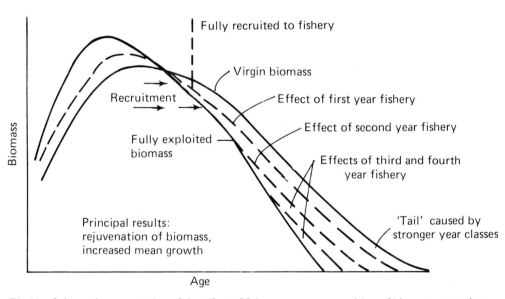

Fig 51 Schematic presentation of the effect of fishery on age composition of biomass, assuming recruitment remains constant and fishing intensity is constant.

133

in balance with spawning stress mortality (*ie* its distribution with age was determined by spawning stress mortality), will decrease until it is in balance with the sum of fishing mortality and spawning stress mortality. This results in the rejuvenation of the population, which has been observed in many fish stocks. The corresponding changes in landings and the catch per unit effort are shown in *Fig 52*. Thus the landings of a given stock which has been taken under exploitation will reach a maximum within a few years, as will the catch per unit effort. Both will decline rapidly thereafter, stabilizing at a lower level. This decline does not mean 'overfishing' as it has been interpreted often in the past. Finally, *Fig 53* illustrates schematically the changes in number and biomass distribution with age if a 'stronger than normal' year class occurs in a given stock.

——Three basic criteria are useful in establishing the procedures for estimation of the Acceptable Catch Estimate (ACE).

(*1*) Maintenance of a reasonably high potential for reproduction of a commercially desirable species, *ie* to keep the biomass at a level where recruitment is not appreciably affected by occasional 'recruitment failures' and especially by too low a spawning biomass. Thus we must know the state of the resource (*ie* the level of the biomass) and consider in addition: age of maturity in relation to the fishery (*ie* fully exploited year-class) and spawning stress mortality, fecundity, spawning period in relation to the

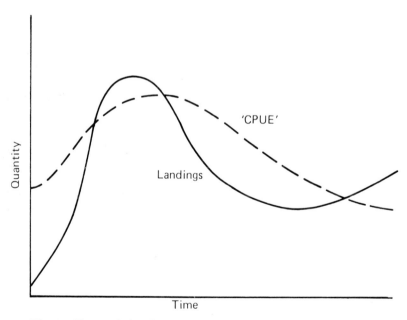

Fig 52 Changes in landings (catches) and 'CPUE' (catch per unit effort) with time when a 'virgin' stock is taken under exploitation.

134

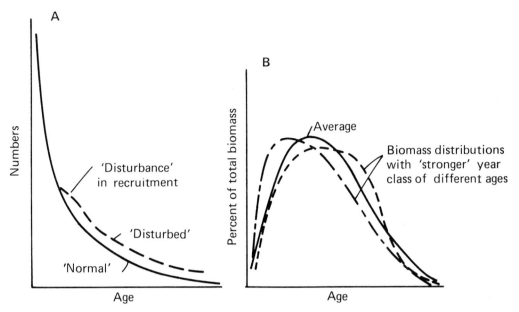

Fig 53 A–Distribution of numbers with age and 'disturbances' in this distribution, and B–average biomass distribution with age and two distributions with stronger year-classes at different ages.

fishery (assuming the fishery can be regulated in space and time), life span of the species, and magnitudes and periods of recruitment variations.

(2) Minimization of any adverse effect on other resources and the ecosystem at large. This requires knowledge of the ecosystem response to the fishery which can be evaluated only in large simulation models. One of the less considered aspects of biomass dynamics is the shrinking of the resource distribution (and spawning areas) with the decrease of biomass. Of concern also are the indirect effects of the fishery on other economically important species (other than target species). The economic aspects enter also into the consideration, and the task of the biologist is to advise what might happen or what is expected to happen.

(3) The harvesting and economic aspects, *ie* maintaining the resource at a level where harvesting is profitable. Although some maximization or optimization of the production would be possible, it is seldom economically and politically possible, *ie* to achieve MSY in an economic sense.
——Additional consideration must be given to the following factors and processes which are expected to influence the estimated acceptable catch.

(4) Can the natural fluctuations which occur in the ecosystem without the influence of fishing be separated from the effects of fishing, and what are the interactions between

fishing and natural fluctuations? The magnitudes and periods of these natural fluctuations caused by a variety of factors, such as temperature anomalies, must be determined with the ecosystem models before introducing high intensity fishing.

(5) What is the process of recovery of a stock, what are the factors determining it, and what are the recovery speeds (relatively sudden or slow and gradual)? Again, the recovery process must be investigated with a full ecosystem model.

(6) What is the state of a given stock in relation to the equilibrium biomass (to be determined from the model)?

——During the iterative procedure in determining the changes in the ecosystem caused by different estimated acceptable catches, additional consideration must be given to the following:

(7) Quantitative changes taking place in principal predator-prey relations in the ecosystem.

(8) Changes in age distribution and recruitment to exploitable stock.

(9) Changes of growth rate and age of maturity.

(10) Changes in bycatch composition in mixed fishery.

The estimation procedure, described below, uses indices, the numerical values of which are derived from past knowledge and experiences as well as from the PROBUB and BIODIS models. Examples of the selected values and their boundaries are given below, together with the explanation of the terms.

The acceptable catch estimate (ACE) is only advisory and used as first-guess input into the ecosystem model. The actually recommended acceptable catches or optimum yields can only be determined by management bodies after the plausible multitude of changes which will take place in the ecosystem are determined in a complete ecosystem model and presented to the management body for evaluation and consideration. The management body, consisting of a variety of interests, must then decide which changes are acceptable in relation to different catches and exploitation strategy.

Indices used in estimating acceptable catch

(i) Direct species indices

B_e ——Exploitable equilibrium biomass (from PROBUB simulation).

B_t ——Actual exploitable biomass present at time t (from PROBUB).

C ——'State of the biomass' index $= B_t/B_e$.

A_s ——Age at which 50% of the population has reached maturity.

A_r ——Age at full recruitment to fishery.

a ——'Spawning biomass saving' index $= A_s - A_r$.
(If > 2, $a = 1$; if 0, $a = 0.7$; if -1, $a = 0.55$; if -2 or smaller, $a = 0.45$).

L ——'Effective' life span (age after which the catches are insignificant; maximum value $A_s + 7$).

M_s ——Spawning stress mortality index $= L - A_s$ (minimum value $= 2$, maximum $= 6$).

g — Growth rate of biomass (in per cent per month of whole biomass, normally between 4·5 and 15).

f — Fecundity index (highly fecund=1·5; very low fecundity=0·9).

M_p — Predation mortality index (indicating the importance of the species as 'forage fish' in the ecosystem; consideration includes prefishery juveniles) $=1+(g/5A_r)$.

r — Recruitment variability index (year-class strength variability; highly variable=3; relatively uniform=2).

n — Index of vulnerability to environmental changes (anomalies) (vulnerable =0·8; least vulnerable=1·0).

h — Index of 'stationarity' of biomass (nonmigratory=1·5; highly migratory=1).

y — Index of fishery in relation to spawning season and area (no or very little fishery in spawning grounds and in spawning season=1·2; fishery mainly in spawning season and on spawning grounds=0·8).

u — Cannibalism index (older species very cannibalistic=1·4; minimal cannibalism=1·0).

(ii) 'Combination Indices'

d — 'Mixed fishery' index (species also caught as bycatch in other fisheries=0·8, species caught mainly in mixed fishery=1·0, species caught mainly as targeted species=1·1).

p — 'Predatory' index (species is important predator on other commercially caught species=0·8, species does not prey on other commercial species to any considerable extent=1·1).

s — 'Prey substitution' index (species can be substituted easily with ecologically and in size/growth similar species=1·1, species substitution possible only from juveniles of dissimilar species=0·9).

The ACE is computed by the following formula, using the indices described above:

$$ACE = \frac{aCB_e}{rM_s} \cdot \frac{M_p fnyu}{h} \cdot \frac{ds}{p} \qquad (71)$$

'Basic ACE'   'Species specific effects'   'Combined effects'

It should be mentioned for general guidance that MSY has varied in the past between 5% and 40% of the exploitable biomass of target species. The ACE of two hypothetical fish – one relatively long-lived demersal species and another relatively short-lived pelagic species – are computed in *Table 18*. The computed ACE's are 19% and 37%, respectively, of the corresponding exploitable biomasses.

Table 18
Examples of estimation of Acceptable Catch Estimate (ACE) of two species. Species 1 is a relatively long-lived demersal fish and species 2 is a relatively short-lived pelagic fish.

Index/subject	Species 1	Species 2
B_e	200,000t	600,000t
B_t	150,000t	660,000t
Exploitable biomass, %	30	65
Exploitable biomass, t	45,000t	429,000t
c	0·75	1·1
A_s	8	2
A_r	7	2
a	1·0	0·7
L	14	5
M_s	6	3
g	4·5	11
f	0·9	1·4
M_p	1·13	2·1
r	2	2·8
n	1·0	0·8
h	1·5	1·0
y	1·0	1·1
u	1·0	1·1
d	0·9	1·1
p	0·9	1·1
s	1·0	1·0
Basic ACE	12,500t	55,000t
Species specific index	0·68	2·85
Combination index	1·0	1·0
ACE	8,500t	156,750t
ACE in % of exploitable biomass	18·9	36·5

16

Use of ecosystem simulation in fisheries management and research

The design of ecosystem simulation (modeling) of a given region already serves at this stage for general research guidance purposes, systematically exposing shortcomings in data and knowledge. Furthermore, the simulation model designs serve also to indicate priorities of research, by suggesting processes of greater importance in terms of greater 'sensitivity' to resultants of determinant processes and those of lesser influence and concern to the ecosystem. Some of the guidance, in respect to need for new emphasis and direction of research, are regional by nature, but many are universal and promote diversity of research as well. Among the examples of new thrust in fisheries research, produced by the ecosystem approach, is the realization of shortcomings in single-species approaches which lack trophodynamic interactions between the species. In contrast to single-species models, one of the most important links in interspecies interactions in ecosystem models is the food relation affecting growth and abundance. Thus, there is a world-wide need to emphasize quantitative fish food studies. Another example is the realization of the necessity to deal with age/size dependent mortalities and to ascertain the predation and spawning stress mortalities in ecosystem models.

Quantitative numerical ecosystem simulations also bring out processes and changes in the ecosystem which have not been easily observed in the past, and permit ecological experiments which would be impossible to conduct in nature. An example is the possible effect of year-to-year variation of predation by squids.

Among the initial scientific uses of the holistic simulations has been the study of the quantitative effects of environmental anomalies, pollutants, and the fishery on the biotic components of the ecosystem. Furthermore, the determination of the carrying capacities of given regions and the study of fluctuations of abundance of species, as caused by various factors internal to the ecosystem, have been successfully demonstrated.

The marine ecosystem simulations, with emphasis on fish, provide new powerful tools for fisheries management in several ways. These simulations not only allow the determination of magnitudes of the resources to be managed and their distributions, but also allow study of space and time variable responses of the ecosystem to any desired and/or prescribed fishery. Ecosystem simulations have shown the necessity to determine the magnitudes and periods of large-scale natural fluctuations in the marine ecosystem

139

which can occur without the influence of the fishery. Without proper evaluation of these fluctuations, the effects of the fishery on the abundance and distribution of the species cannot be evaluated. The simulations are used for the evaluation of the effect of various species composition of total catches (catch of all species from a given region), either present or prospective, on the ecosystem at large. Each given species composition of total catch would result in a different ecosystem in a number of years, whereby major changes will take place within four to six years. Superimposed on these changes caused by fishery will be natural fluctuations caused by a variety of factors.

The marine ecosystem is quite unstable in space and time. Thus it is difficult to obtain high reliability in any prognostic calculations. The great uncertainties in such prognoses are still found in the great fluctuations in recruitment and in juvenile mortalities. Present knowledge on the whereabouts of prefishery juveniles is also very deficient in most species.

One of the main uses of ecosystem simulations in fisheries management and research is in resource evaluation as illustrated by numerous examples in this book. On this subject direct validation of the simulation is provided through various resource surveys (see Chapter 13).

The ecosystem simulation models provide information of average rates of recoveries and rates of changes of biomasses in general. This information has not been available from conventional fisheries research (Chapter 11).

Fishing effort in relation to catch is a density-dependent process. Consequently, a given constant total catch requires different efforts at different biomass levels. Management decisions for regulation of the fishery by total allowable catch require the best available and reliable knowledge of the level of biomass for the species under consideration. Various biomass levels for this purpose can be simulated with the DYNUMES model which can be used in a relative and comparative manner to estimate the effects of fishery. The fishery can also be simulated as a spatial and temporal variable according to realistic possibilities and their limits.

The principles of the application of ecosystem simulation to the effects of the fishery have been described in Chapter 14. It was shown that ecosystem simulation will provide a quantitative picture of the state of the marine ecosystem with different levels of the fishery on any species. It will also provide a means to estimate how to effect changes in the ecosystem with a managed fishery – *eg* how and to what extent the resource levels can be manipulated and what are the rates of changes (including the recovery time).

Although the fishing industry (both catching and processing) will largely determine the fisheries management and agree on which changes in the ecosystem are acceptable from an economic point of view, the fisheries scientists must advise on the desirable state of the ecosystem from the point of view of its 'health'.

At present very few general guidelines can be established for the judgment of the state of health of marine ecosystems. Much more research is needed to establish principles and guidelines to this end and to gain experiences on the changeability of the biota in the marine ecosystem. One of the main tools in this research will be the numerical marine ecosystem simulation models.

The definition of the 'desirable' or acceptable state of the marine ecosystem is dependent on characteristics of each region. It is, therefore, difficult or nearly impossible to establish universally applicable criteria for this purpose. Even general statements such as 'it might be desirable to fish down a less valuable predator to increase the biomass or productivity of other more valuable species', are not universally accepted.

In addition to questions pertaining to biomass resource levels, there are several important management questions pertaining to processes in the ecosystem which have not been discussed in this book, partly due to lack of information. One of these questions is, 'How large a spawning biomass is necessary for any given species in a given region to ascertain adequate recruitment?' This problem can also, at least in essence and principle, be investigated by ecosystem simulation models with additional specific subroutines.

In the past the biologists have focussed on fish yield, the economists have considered cost/price ratios, and the politicians have concentrated on minimization of conflicts, which might arise from different interests and desires of fishermen, processors, and investors as well as different nations. Often the fisheries management problem is not a biological, but a political one.

In the past it has been difficult to evaluate the performance of fishery management. The ecosystem simulation approach would alleviate this task considerably. Ecosystem simulation allows the determination of management options before they are put into practice, and allows the continual monitoring of the system. Thus, we can evaluate the impact of the management and determine whether the management measure is making any difference, and, if so, what, and we can match the fishing effort to the conditions of the stocks.

The applications of ecosystem simulations are indeed numerous and far from being fully explored and utilized. We might visualize these seemingly unlimited possibilities if we consider that in essence we are simulating nature (*ie* the ecosystem) and its functioning quantitatively by computer and, to the extent that we have reproduced the ecosystem, can review the whole system of unseen natural processes at our desks.

17

List of symbols

a	——empirical constant (also food coefficient for growth).
a_n	——fraction of a year class of total exploited population.
A	——constant.
A_i	——age of species i.
A_g	——half of the annual range of growth coefficient change.
A_1	——half-range of annual main (spring) plankton maximum.
A_2	——half-range of annual secondary (fall) plankton maximum.
ACE	——'acceptable catch estimate'.
b	——food required for maintenance in terms of fraction of body weight daily.
B	——biomass.
B_a	——biomass of apex predator a.
$B_{e,i}$	——equilibrium biomass.
B_b	——initial biomass (analogue to equilibrium biomass).
$B_{i(t)}$	——biomass of species i, in time step t.
B_n	——biomass of year class n.
B_{n+1}	——biomass in next older year class.
B_{pn}	——per cent fraction of biomass of year class n.
B_o	——virgin biomass.
c	——food required for sex products development in terms of fraction of body weight.
C	——consumption (predation mortality).
C_a	——food requirement of apex predator a.
C_i	——consumption of species i.
$C_{i,j}$	——consumption of species i by species j.
$C_{i,a}$	——consumption of species i by apex predator a.
D	——constant.
e	——base of natural logarithms.
f	——variable feeding level (also function).
f_a	——fraction of spawning (exploitable) biomass of total biomass.
f_n	——mean increase factor of numbers of fish in next older year-class.

F	——fishing mortality (coefficient).
F_n	——changed (new) fishing mortality coefficient.
F_0	——original (old) fishing mortality coefficient.
g	——growth coefficient.
g_c	——growth coefficient compensated to recruitment.
g_m	——half of the magnitude of annual change of growth coefficient.
g_n	——instantaneous growth rate of a year-class.
g_0	——basic growth coefficient (annual mean).
g_t	——growth coefficient at time t.
G	——growth of biomass.
G_{an}	——annual growth rate (percentage) of year class n.
G_{mn}	——monthly growth rate (percentage) of year-class n.
h	——coefficient.
H	——food requirement for growth parameter.
i	——species indicator.
j	——species indicator.
k	——iteration constant.
k_m	——food requirement for metabolism parameter.
k_0	——fraction of biomass removed from a given grid point by 'forced migration'.
k_s	——fraction of biomass added to a grid point by 'forced migration'.
K	——constant.
l	——length of the grid.
m	——grid coordinate.
m	——mortality coefficient (from old age and diseases).
M	——natural mortality (also natural mortality coefficient).
M_n	——natural mortality coefficient.
M_p	——predation mortality.
M_s	——spawning stress mortality.
MSY	——maximum sustainable yield.
n	——grid coordinate.
n_s	——number of years after the cohort has reached 80% maturity.
N	——number of fish.
N_n	——number of fish in year-class n.
N_{n+1}	——number in next older year-class.
$p_{i,a}$	——decimal fraction of species i in the food of apex predator a.
$p_{i,j}$	——fraction of species j in the food of species i.
$p_{i,j,t}$	——decimal fraction of food item j in the food of species i at time t.
$p_{i,j,0}$	——same as $P_{i,j,t}$ except for time 0 (prescribed).
p_m	——fraction of food missing from full ration.
p_{mj}	——changed fraction of food item.

p_{vj}	——unchanged fraction of food item.
$P(P_1, P_2)$	——population (biomass), (also stock).
P_m	——sum of the fraction of food missing from full ration.
P_o	——annual mean plankton standing stock.
P_s	——unchanged food composition.
P_t	——plankton standing stock at time t.
q_a	——food requirement of apex predator a (in percentage of body weight daily).
$q_1...q_n$	——food requirements of species i to n.
Q	——biomass removal (total mortality and fishery).
Q_f	——food availability.
$r_{g,i}$	——food requirement for growth (ratio of growth to food required for growth)
r_i	——fraction of body weight required daily for maintenance.
R	——accumulating food consumed (also total food requirement).
R_i	——ration (normal food requirement).
R_o	——maximum fraction of food item allowed to be consumed.
R_r	——recruitment.
R_{t-1}	——fraction of food item consumed in previous time step.
S	——starvation in terms of fraction of missing food.
S_i	——shortage of food (starvation).
S_n	——starvation in terms of quantity of food missing.
t	——time, [(t)-time step t; (t-l) – previous time step]
t_d	——time in days (also length of time step).
T	——temperature.
T_c	——computed biomass turnover rate.
T_o	——optimum (acclimatization) temperature.
U	——u component of migration speed.
UT	——'upmigration' gradient of biomass (u direction).
V	——v component of migration speed.
V_{max}	——maximum migration speed component.
V_p	——fraction of biomass allowed to be taken from a given species at a given time step, in addition already taken by other species.
V_x, V_y, V_z	——migration speed in x, y and z direction.
VT	——'upmigration' gradient of biomass (v direction).
$W(w)$	——weight (body weight).
W_n	——mean weight of specimen in year-class n.
W_{n+1}	——mean weight of specimen in next older year-class.
x	——space coordinate.
y	——space coordinate.
Y	——yield.
z	——space coordinate.
Z	——total mortality (also total mortality coefficient).

Z_n	——total mortality (in biomass) of year-class n.
γ	——smoothing coefficient (0·8 to 0·96).
α	——phase speed.
β	——smoothing coefficient $[=(1-\gamma/4)]$.
α_1,α_2	——phase speeds.
ϕ_i	——fishing mortality coefficient.
κ_g	——phase lag of annual growth coefficient change.
κ_1,κ_2	——phase lags (month of annual maximum).

18

References

AKIMUSHKIN, I I. Golovonogie mollyuski morei
1963   SSSR (Cephalopods of the seas of the
        USSR). *Akad. Nauk. SSSR, Inst.
        Okeanol*, Moskva. In Russian. (Transl.
        by Israel Prog. Sci. Transl, 1965, 223p,
        avail. Natl. Tech. Inf. Serv, Springfield,
        VA as TT 65–50013.)

ALEXANDER, V. Sea ice and primary production
1978   in the Bering Sea. Univ. of Alaska,
        Fairbanks, MS report.

ALTON, M S.Bering Sea benthos as food
1973   resource for demersal fish populations.
        In Oceanography of the Bering Sea with
        emphasis on renewable resources.
        Eds. D W Hood and E J Kelley. *Occ.
        Publ. 2, Inst. Mar. Sci, Univ. Alaska*,
        257–279.

ALVERSON, D L. A study of annual and
1960   seasonal bathymetric catch patterns for
        commercially important groundfishes of
        the Pacific northwest coast of North
        America. *Pac. Mar. Fish. Comm.
        Bull. 4.*

ANDERSEN, K P and URSIN, E. A multispecies
1977   extension to the Beverton and Holt
        theory of fishing, with accounts of
        phosphorus circulation and primary
        production. *Meddr. Danm. Fisk.–og
        Havunders.* N.S. 7:319–435.

BAKKALA, R G and SMITH, G B. Demersal fish
1978   resources of the eastern Bering Sea:
        Spring 1976. Natl. Mar. Fish. Serv,
        Northwest and Alaska Fish Cen.,
        Seattle, Wash., Processed Rpt., 233p.

BALCHEN, J G. Modeling and identification of
1980   marine ecological systems with
        applications in management of fish
        resources and planning of fisheries

operations. Modeling, Identification
and Control. Norw. Res. Council for
Scient. and Industr. Res. 1(2):67–68.

BARANOV, F I. On the question of the
1918   biological basis of fisheries. *Nauchnyi
        issledovatelskii ikhtiologicheskii Institut.
        Izvestiia.* 1(1):81–128.

BARANOV, F I. On the question of the dynamics
1926   of the fishing industry. *Bull. Rybnovo
        Khoziaistva*, 1925. (8):7–11.

BEHRENS, J Chr. A semi-analytical sensitivity
1978   analysis of non-linear systems. Inst.
        of Mathem. Stat. and Oper. Res.,
        Techn. Univ. Denmark, Res. Rpt.
        4/1978:29pp.

BEST, E A. Halibut ecology. *In* Fisheries
1979   Oceanography – Eastern Bering Sea
        Shelf. Northwest and Alaska Fish.
        Cen., Seattle, Wash., Processed Rpt.
        79–20:127–165.

BEVERTON, R J H. Maturation, growth, and
1963   mortality of clupeid and engraulid
        stocks in relation to fishery. *Rapp.
        P–V. Réun. Cons. Int. Explor. Mer*,
        154:445–67.

BEVERTON, R J H and HOLT, S J. On the
1957   dynamics of exploited fish populations.
        Fishery Invest. London, Ser.
        2(19):533pp.

COLEBROOK, J M. Changes in the zooplankton
1978   of the North Sea. *Rapp. P–V. Réun.
        Cons. int. Explor. Mer*, 172:390–396.

CUSHING, D H. *Fisheries biology. A study of
1968   population dynamics.* Univ. of Wisconsin
        Press, Madison, Milwaukee and
        London. 200pp.

CUSHING, D H. Recruitment and parent stock
1976   in fisheries. Univ. Wash., Div. Mar.

Resour., Wash. Sea Grant Program,
WSG, 73–1, 197pp.

DAAN, N. A quantitative analysis of the food
1973   intake of North Sea cod, *Gadus
morhua. Neth. J. Sea Res.* 6(4):479–517.

DAAN, N. Some preliminary investigations into
1976   predation on fish eggs and larvae in the
southern North Sea. ICES CM
1976/L:15, Plankton Committee.

DEMENTJEVA, T F. Correlations between
1964   indices of relative abundance of young
fish, recruitment size and maturity rate
as a basis for annual prediction. *Rapp.
P–V. Réun. Cons. int. Explor. Mer,*
155:183–187.

DICKIE, L M. Predator-prey models for
1979   fisheries management. *Predator-prey
systems in fisheries management,* Sport
Fishing Inst., Washington, DC:281–292.

FLÜCHTER, J and TROMMSDORF H. Nutritive
1974   simulation of spawning in common sole
(*Solea solea* L) *Ber. Dtch. Wiss. Komm.
Meeresforsch.* 23:352–359.

GRAINGER, R J R. Herring abundance off the
1978   west of Ireland in relation to
oceanographic variation. *J. Cons. int.
Explor. Mer,* 38(2):180–188.

GROSSLEIN, M D. Some results of fish surveys
1976   in the mid-Atlantic Bight, important for
assessing environmental impacts. *Am.
Soc. Limnol. and Oceanogr. Spec.
Symp.* 2:312–328.

HALVER, J E (Ed). Fish nutrition. Academic
1972   Press, NY and London.

HARDEN JONES, F R, ARNOLD G P, GREER
1979   WALKER, M, and SCHOLES P. Selective
tidal stream transport and the migration of plaice (*Pleuronectes platessa* L)
in the southern North Sea. *J. Cons. int.
Explor. Mer,* 38(3):331–337.

HARRIS, J G K. The effect of density-dependent
1975   mortality on the shape of the stock and
recruitment curve. *J. Cons. int. Explor.
Mer,* 36(2):144–149.

HEINRICH, A K. On the production of copepods
1962   in the Bering Sea. *Int. Revue ges.
Hydrobiol.* 47(3):465–469.

ISAACS, J D. Some ideas and frustrations about
1976   fishery science. *Cal. Coop. Oceanic Fish
Invest. Rpts.,* Vol. 18:345–43.

JAKOBSON, J. Exploitation of the Icelandic
1978   spring and summer spawning herring in

relation to fisheries management
1947–1977. ICES Symp. on Biol.
Basis of Pelagic Fish Stock Management.
Paper 2 (in press in *Rapp. P–V. Réun.
Cons. int. Explor. Mer*).

JONES, R and HISLOP, J R G. Further
1978   observations on the relation between
food intake and growth of gadoids in
captivity. *J. Cons. int. Explor. Mer,*
38(2):244–251.

KOTO, H and MAEDA, T. On the movement of
1965   fish shoals and the change of bottom
temperature on the trawl-fishing ground
of the eastern Bering Sea. *Bull. Jap.
Soc. of Scientific Fisheries* 31(1):769–774.

KREMER, T N. and NIXON, S W. *A coastal
1977   marine ecosystem. Simulation and
analysis.* Springer-Verlag, NY. 217pp.

KRUEGER, F. Neuere matematische
1964   Formulierung der biologischen
Temperatur-funktion und des
Wachstums. *Helolander Wiss.
Meeresunters.* 9, 108–124.

LAEVASTU, T and FAVORITE, F. Dynamics of
1976   pollock and herring biomasses in the
eastern Bering Sea. Northwest and
Alaska Fish. Cen., Seattle, Wash.,
Proc. Rpt. 18pp.

LAEVASTU, T, DUNN, J and FAVORITE, F.
1976   Consumption of copepods and
euphausiids in the eastern Bering Sea as
revealed by a numerical ecosystem
model. Paper L:34, ICES CM 1976,
Plankton Comm. 10pp.

LAEVASTU, T and FAVORITE, F. Numerical
1978a   evaluation of marine ecosystems. Part I.
Deterministic Bulk Biomass Model
(BBM). Northwest and Alaska Fish.
Cen., Seattle, Wash., Proc. Rpt., 22pp.

LAEVASTU, T and FAVORITE, F. Numerical
1978b   evaluation of marine ecosystem. Part II.
Dynamical Numerical Marine
Ecosystem Model (DYNUMES III) for
evaluation of fishery resources.
Northwest and Alaska Fish. Cen.,
Seattle, Wash., Proc. Rpt., 29pp.

LAEVASTU, T and FAVORITE, F. Fish biomass
1978c   parameter estimations. Northwest and
Alaska Fish. Cen., Seattle, Wash.,
Proc., Rpt., 16pp.

LAEVASTU, T and FAVORITE, F. Holistic
1981   simulation of marine ecosystem. *In*

A R Longhurst (ed), Analysis of Marine Ecosystems, pp702–727. Academic Press, Inc., London.

LAEVASTU, T, LIVINGSTON, P and NIGGOL, K.
1980   Basic inputs to PROBUB model for the eastern Bering Sea and western Gulf of Alaska. Northwest and Alaska Fish. Cen., Seattle, Wash., Proc. Rpt. 80–3.

LARKIN, P A. An epitaph for the concept of
1977   maximum sustained yield. *Trans. Am. Fish. Soc.* 106(1):1–11.

LAST, J M. The food of larval turbot,
1979   *Scophthalmus maximum* L, from the west central North Sea. *J. Cons. int. Explor. Mer*, 38(3):308–313.

LETT, P G and KOHLER, A C. Recruitment: a
1976   problem of multispecies interactions and environmental perturbations, with special reference to Gulf of St. Lawrence Atlantic herring (*Clupea harengus harengus*). *J. Fish. Res. Board Can.* 33:1353–1371.

LONGHURST, A R. Ecological models in
1978   estuarine management. *Ocean Management* 4 (1978):287–302.

LONGHURST, A R and RADFORD, P J.
1975   GEMBASE I. Internal documents. Inst. for Mar. Env. Res., Plymouth.

MITO, K. Food relationships among benthic
1972   fish populations in the Bering Sea. Hokkaido Univ. Grad. School, Hakodate, Japan, MS Thesis, 135pp.

MOTODA, S and MINODA, T. Plankton of the
1974   Bering Sea. *In* Oceanography of the Bering Sea with emphasis on renewable resources. Eds. D W Hood and E J Kelly, *Occ. Publ. No. 2, Inst. Mar. Science, Univ. Alaska*, 207–230.

PALOHEIMO, J E and DICKIE, L M. Food and
1965   growth of fishes. I. A growth curve derived from experimental data. *J. Fish. Res. Board, Canada*, 22(2):521–542.

RADOVICH, J. Catch-per-unit-effort: fact,
1976   fiction or dogma. *Cal. Coop. Oceanic Fish Invest Rpts.*, 18:31–33

RICKER, W E and FOERSTER, R E. Computation
1948   of fish population. A symposium on fish populations. *Bull. Bingham Oceanogr. Coll.* 11(4):173–211.

ROTHSCHILD, B J and FORNEY, J L. The
1979   symposium summarized. *In* R H Stroud (compiler) and H Clapper (editor), *Predator-prey systems in fisheries management*, pp487–496. Sport Fish. Inst., Washington, DC.

SHABONEEV, I E. O biologii i promysle sel'di
1965   vostochnoi chasti Beringova morya (Biology and fishing of herring in the eastern part of the Bering Sea). *Tr. Vses. Nauchno-issled. Inst. Morsk. Rybn. Khoz. Okeanogr. 58 (Izv. Tikhookean. Nauchno-issled. Inst. Morsk. Rybn. Khoz. Okeanogr. 53)*:139–154. In Russian. (Transl. by Israel Prog. Sci. Transl., 1968. p 130–146 *in* P A Moiseev (ed), Soviet fisheries investigations in the northeast Pacific, Pt. 4, avail. Natl. Tech. Inf. Serv., Springfield, Va. as TT 67–51206.)

SMETANIN, D A. On the evaluation of the
1956   organic production in several areas of the Bering and Okhotsk Seas. *Tr. Inst. Okeanol.* 17.

TYLER, A V and DUNN, R S. Ration, growth
1976   and measures of somatic and organ condition in relation to meal frequency in winter flounder, *Pseudopleuronectes americanus*, with hypothesis regarding population homeostasis. *J. Fish. Res. Board, Canada*, 33(1):63–75.

URSIN, E. Single species and multispecies fish
1979   stock assessment. MS presented at 16th Nordic Fisheries Conference, Marrehamn, Finland.

148

19

Explanation of terms

ABC	Acceptable (also allowable) Biological Catch. Subjectively estimated amount of catch of given species from a given region. Has no objective scientific definition.
Abiotic factors	Physical and chemical environmental factors, such as temperature, salinity, depth, *etc*, which affect the abundance and distribution of plants and animals.
Acclimatization	is the phenomenon through which an organism becomes habituated to a climate not native. The same term is used also when referring to an 'aquatic climate', that is, to environmental conditions in the sea.
Acclimatization temperature	Annual mean temperature (also range of temperature) of the normal habitat of a species or a fish stock.
ACE	Acceptable Catch Estimate. A preliminary estimate of catch of a given species which might be taken from a stock in a given region (see Chapter 15 for methods of its estimation).
Advection	or advective water movement means basically horizontal transport of waters (by currents).
Age-class	Specimens of a given species which are approximately of the same age (*ie* born in the same spawning season).
Anabolism	The process in plants and animals by which food is changed into living tissue; constructive metabolism.
Apex predators	Those predators in an ecosystem which are normally not preyed by others in this system (*eg* marine mammals, birds, sharks, *etc*).
Baroclinic	Condition in the ocean and in the atmosphere where constant pressure surfaces intersect constant density surfaces.
Barotrophic	Assumed (simplified) condition in the atmosphere where constant pressure and constant density surfaces are identical.
Benthos	Animals and plants which live on or in the bottom, and also those whose life is in some way closely connected with the bottom.
BIODIS	A computer programme in NWAFC, used for computation of

	various parameters of a fish population, such as growth rates and biomass distribution with age.
Biomass	is the wet weight or mass of living organisms.
Biomass-based model	Any single-species, multispecies, or ecosystem model (simulation) which uses biomass of the age class and/or stock as the basic measure (in contrast to number-based models, which use numbers of specimens for the same purpose).
Biomass distribution	Quantitative distribution of biomass with age in a given stock.
Biota	refers to all fauna and flora.
Biotic factors	Factors of a biological nature, such as availability of food, competition between species, predator and prey relationships, *etc*, which, besides the purely environmental factors, also affect the distribution and abundance of a given species of plant or animal.
Buffering	Absorbing, cushioning, and/or lessening the impact of a change, shock, or conflict.
BWD	Body Weight Daily. A measure of food requirement and/or uptake (expressed in percent or in decimal fraction).
Caloric value	Heat value (or energy content) of organisms and their food. (Often not well defined and expressed either on the basis of dry weight or biomass.)
Carrying capacity	The amount of biota (*eg* finfish) which a defined ocean area can sustain.
Catabolism	Destructive metabolism, destruction of living matter.
Cohort	Age-class of a species (fish).
Consumption	Uptake (use) of food.
Continental shelf	The sea (bottom) from about 200m depth to coast.
Continental slope	The sea (bottom) off the continental shelf where depth increases rapidly with distance towards open ocean (usually 200m to 500m depth zone).
Deep water (area)	Ocean areas (open ocean) where depth is greater than 500m.
Demersal	means 'living near the bottom' (from the Latin *demersus*, meaning 'plunged under').
Density dependent	Indicates that a result of a process or of a state is dependent on the density of a species present [*ie* abundance in a unit area or volume of the subject (*eg* fish, its food, *etc*)]. A general and widely used term in fisheries biology, but not well defined.
Diagnostic phase	Analytical phase in a forecasting process. The distributions of subjects or parameters are determined for a given (sypnotic) time in the diagnostic phase.
DYNUMES	Dynamical Numerical Marine Ecosystem Simulation (model).

Ecosystem	Plants and animals in a specified environment, depending on and interacting with this environment and with each other.
Equilibrium biomass	Biomass of a given species which can be sustained in a given region under defined (balanced) conditions.
Equilibrium state	Balanced state, state of no change.
Expoitable biomass	The biomass of fishable size fish in a given stock.
EY	Equilibrium Yield. Subjectively estimated amount of catch which could be taken from a given region without changing the biomass. Has no objective scientific definition.
Fecundity	The quality of producing offspring. It is measured in fish by the average number of eggs produced by a female.
First guess	A subjectively estimated (plausible) preliminary value of a parameter. Will usually be recomputed in a model and/or simulation.
Flow diagram	A diagram showing the types and sequence of computations in a computer programme.
Food chain	The sequence of using plants and animals as food. The food chain starts with plants (primary producers), followed by plant-eating animals (herbivores), thirdly come animals who feed on herbivores (first stage carnivores), *etc*. However, this sequence does not occur in nature in this pure form, as the food composition of most animals varies in space and time and also changes with age (size) of the organisms.
Food pyramid	This form is quasi-synonymous to food chain. However, food pyramid attempts to present a quantitative picture of the abundance of different links in the food chain: on the bottom of the pyramid is the primary organic production and on top of the pyramid are the apex predators. The same limitations apply to food pyramid as to food chain.
Food requirement	Annual mean amount required by the biomass of given species to produce mean growth. Food requirement is synonymous to mean ration. Food requirement is usually expressed as amount of food biomass required by unit biomass of organism in terms of BWD (body weight daily). Food requirement is also separated into different components such as food requirement for growth, food requirement for maintenance, *etc.*
Forced migration	Migration of fish in or out of a defined area, caused by *eg* environmental anomalies (*eg* too cold or too warm) or pronounced events, such as storms.
Grid	in numerical computation is a net of squares (and/or quadrangles) drawn over an area where computation is made at each grid intersection.

Holistic ecosystem (simulation)	Numerical simulation of (marine) ecosystem which contains all biota, environment and all essential processes within this ecosystem.
Individual time change	Quantitative change of a property (such as temperature) or a quantity (such as a given biomass) with time in a specified location. Individual time change consists of local time change and advection.
Juveniles	Juveniles are comprised of young fish after larval stage to the age (size) they become subject to fishing or become sexually mature.
Local time change	Change of a property (such as temperature) or a quantity (such as a biomass) with time, caused by various processes in this location, such as local cooling or biomass growth and/or mortality.
Maintenance (sustenance) ration	Average amount of food required by a unit fish biomass to maintain the body weight and activity, but allowing no growth nor decrease of weight.
Metabolism	meaning actually a 'change', refers to chemical changes taking place within living material (reactions within an organism).
MSY	Maximum Sustainable Yield. A concept in earlier fisheries management which cannot be objectively defined, except with unreal assumptions.
Natural mortality	Mortality of fish from other causes than fishing. Natural mortality includes predation mortality, senescent mortality, spawning stress mortality, and mortality from diseases.
Nekton	Mobile aquatic animals which have the ability (in spite of currents and waves) to actively determine their course of movement.
Normalization	A mathematical procedure where a given quantity is usually made unity (*ie* equated to one) and subquantities are expressed as fraction (or percentage) of this unity. Normalization to any other quantity than one can also be made.
Number-based model	A single or multispecies model where main computation procedures are based on numbers of fish in a given population or in a given year-class.
Numerical model	Computerized models where inputs and outputs are numerical values (quantities) and all computations are done numerically.
Nutrients	in sea water refers to elements required to support the growth of phytoplankton in the sea. These include usually phosphates, nitrates, and silicates, but sometimes also the minor elements of sea water, such as copper, manganese, cobalt, iron, *etc*, are included.
Open boundary	Boundaries in two or three dimensional numerical models, through which quantities are lost or gained (*eg* migrations through boundaries).
OY	Optimum Yield. In reality undefinable quantity, used in some

	fisheries management considerations.
Parameter	of a mathematical function is a quantity to which the operator may assign any arbitrary value, as distinguished from a variable, which can assume only those values that the function makes possible.
Pass (in numerical operation)	Application of a given numerical operation on a numerical array. This is normally done with DO loops in FORTRAN programme. Operators (such as smoothing parameters) can be changed in different passes.
Pelagic	refers to all organisms living freely in the water masses. Analogously the mid-water trawl is called pelagic trawl.
Phase lag	A term used in connection with harmonic formulas. Phase lag (given usually in degrees or in radians) determines the maximum (and minimum) in a time-dependent harmonic formula.
Phase speed	Length of time step in a harmonic formula, usually given in degrees or in radians.
Population dynamics	Quantitative changes occurring in a population with time. The term is often applied to signify mathematical approaches to simulate the dynamics of single-species populations.
Predation	Catching and eating of other animals. Often this term is used as a synonym to grazing.
Prerecruit	Juvenile fish (age-classes) which are not yet fully recruited to fishery. This term is often used as a synonym to juveniles.
Primary production	Production of organic matter by plants (mainly phytoplankton in the sea).
Primitive equation (PE) model	Numerical models which use basic equation of motion and equation of state.
PROBUB	Prognostic Bulk Biomass model.
Prognostic phase	Forecasting phase in any time-dependent model.
Ration	Synonym to food requirement (in this book). Annual mean amount of food required by a given unit amount of biomass.
Recovery of stock	Process of increase of the biomass of a given species from a previous lower level of abundance after decrease of fish effort. This term is often applied to exploitable part of the biomass, and is then dependent on the recruitment from younger year-classes.
Recruitment	Process of movement of age (size) classes towards older age-classes measured quantitatively by the amounts of the species biomass moving up to older age classes per unit time (usually in a year).
Recruitment overfishing	If the amount of fishing will decrease considerably the spawning biomass (or the number of spawners), recruitment overfishing is said to occur. This condition has never been quantitatively defined.

Regeneration of nutrients	Break-down of organic matter (*eg* by bacteria) to chemical components (mainly inorganic).
Rejuvenation of population	Decrease of average age of a given population, either by the decrease of exploitable biomass or increase of juvenile biomass due to good recruitment.
Roundfish	Semi-demersal species, mainly gadids. Term used mainly in Europe.
Secondary production	Production of organic matter by conversion of eaten organic matter into new living tissue.
Section	A two-dimensional picture of distribution of properties and/or biota. Also a line of observation and/or collecting stations.
Semipelagic (semidemersal)	Fish who spend part of their life on the bottom and part in the water mass above and obtain their food from both regimes.
Senescent mortality	Mortality due to old age.
Sensitivity analyses	Determination of the numerical response of a function to the change of the value of any parameter of this function.
Simulation	Quantitative (numerical) reproduction of marine ecosystem, based on all available empirical knowledge of this system.
Sink (of biomass)	Decrease of biomass of a given species (*ie* mortality exceeds growth).
Smoothing	Averaging of numerical values in an array presenting area or time sequence.
Source (of biomass)	Increase of biomass of a given species (*ie* growth exceeds mortality).
Spawning stress mortality	Mortality during and shortly after spawning.
Stability criterion	Criterion for the length of maximum time step in some formulas which are solved in finite difference form.
Standing stock	Size of the stock of a given species (in numbers or in biomass) in a given region.
Synoptic	means simultaneous. In the daily weather service the weather observations made at various places for obvious reasons must be simultaneous. Thus 'actual meteorology' has been given the name 'synoptic meteorology'. Analogously 'actual oceanography', eventually leading to an information service on the actual phenomena in the oceans, must be based on more or less synoptic data. Therefore, synoptic oceanography or synoptic hydrography is similarly called 'hydropsis'.
Target species	Species which is main subject for fishing by a given vessel with a given type of gear.

154

Trophic level	Level or link in food chain, *eg* herbivores, first stage carnivores, *etc*. Due to great variations in food composition, these levels cannot be defined with any accuracy in an ecosystem.
Trophodynamics	Dynamics of feeding and quantitative considerations of conversion of food to tissue and for activity (maintenance).
Time step	Time interval of computations in a time-dependent model.
U component	Horizontal or east-west component of velocity.
Unique solution	Only solution possible, satisfying a set of equations with defined input data (parameters).
Upcurrent method	A method for computation of change of a property or a subject in a model with two-dimensional horizontal grid, where the gradient of the property from the upcurrent direction is advected towards the grid point under computation.
Validation	Comparison of results of a simulation (model) with empirically obtained data (*ie* with direct measurements).
V component	Vertical or north-south component of velocity.
Verification	Checking of individual formulas and logic in a simulation (model) with best available empirical knowledge and tested theory.
VPA	Virtual Population Analysis.
Year-class (strength)	Numbers or biomass of a given species born in a given year.
3-D space	Space presentation in a simulation with two horizontal dimensions and one vertical (depth) dimension.

20
Subject index

157

Other books published by
Fishing News Books Ltd
Free catalogue available on request

Advances in aquaculture
Advances in fish science and technology
Aquaculture practices in Taiwan
Atlantic salmon: its future
Better angling with simple science
British freshwater fishes
Commercial fishing methods
Control of fish quality
Culture of bivalve molluscs
Echo sounding and sonar for fishing
The edible crab and its fishery in British waters
Eel capture, culture, processing and marketing
Eel culture
European inland water fish: a multilingual catalogue
FAO catalogue of fishing gear designs
FAO catalogue of small scale fishing gear
FAO investigates ferro-cement fishing craft
Farming the edge of the sea
Fish and shellfish farming in coastal waters
Fish catching methods of the world
Fisheries of Australia
Fisheries oceanography and ecology
Fishermen's handbook
Fishery products
Fishing boats and their equipment
Fishing boats of the world 1
Fishing boats of the world 2
Fishing boats of the world 3
The fishing cadet's handbook
Fishing ports and markets

Fishing with electricity
Fishing with light
Freezing and irradiation of fish
Handbook of trout and salmon diseases
Handy medical guide for seafarers
How to make and set nets
Inshore fishing: its skills, risks, rewards
Introduction to fishery by-products
The lemon sole
A living from lobsters
Marine pollution and sea life
The marketing of shellfish
Mending of fishing nets
Modern deep sea trawling gear
Modern fishing gear of the world 1
Modern fishing gear of the world 2
Modern fishing gear of the world 3
More Scottish fishing craft and their work
Multilingual dictionary of fish and fish products
Navigation primer for fishermen
Netting materials for fishing gear
Pair trawling and pair seining: the technology of two boat fishing
Pelagic and semi-pelagic trawling gear
Planning of aquaculture development: an introductory guide
Power transmission and automation for ships and submersibles
Refrigeration on fishing vessels
Salmon and trout farming in Norway
Salmon fisheries of Scotland
Scallops and the diver-fisherman
Seafood fishing for amateur and professional
Seine fishing: bottom fishing with rope warps and wing trawls
Stability and trim of fishing vessels
Study of the sea
The stern trawler
Textbook of fish culture: breeding and cultivation of fish
Training fishermen at sea
Trout farming manual
Tuna: distribution and migration
Tuna fishing with pole and line